Burt, D R R

Platyhelminthes and parasitism; an introduction to parasitology [by] D. R. R. Burt. New York, American Elsevier Pub. Co., 1970.

viii, 150 p. illus., plates. 23 cm. ([Modern biology series]) 8.75

Bibliography: p. [138]-140.

1. Platyhelminthes. 2. Parasitism. I. Title.

QL391.P7B86 595'.12 77-104784

SBN 444-19697-8 MARC

PLATYHELMINTHES AND PARASITISM

PLATYHELMINTHES AND PARASITISM

An Introduction to Parasitology

D. R. R. BURT, B.Sc., F.L.S., F.R.S.E.

Formerly Professor of Zoology,
University of Ceylon, and
Senior Lecturer in Zoology,
University of St Andrews

AMERICAN ELSEVIER PUBLISHING COMPANY, INC.

NEW YORK 1970

American Elsevier Publishing Company, Inc.
52 Vanderbilt Avenue, New York, New York 100 17

Standard Book Number 444-19697-8
Library of Congress Catalog Card Number 77-104784

PRINTED IN GREAT BRITAIN

Preface

The animals that we know as flatworms constitute one of the great groups or phyla in the animal kingdom. Within this group there are independent, free-living forms and forms that are completely dependent on others, while between these extremes there is a bewildering variety of different kinds and degrees of association. This book is mainly an account of Platyhelminthes intended as an introduction to the study of that fascinating subject parasitism, and it is written primarily for student biologists.

With the exception of the more common parasites of man and his domestic animals, the vast majority of Platyhelminthes are known only by their scientific names, and few of these are to be found in elementary text-books. The reader is accordingly faced with a number of scientific names of animals which, in the absence of a framework, must be more or less meaningless. A chapter has therefore been included on the classification of the phylum giving the characters of the various classes and many of the orders, exemplified by the animals referred to in the text. It is hoped that this will ease the difficulty that the beginner must feel and it is suggested that this chapter be used more or less as a ready reference to orientate the animals discussed later in the book. A classification can only be appreciated after one has a knowledge of the animals concerned.

The various associations that are met with between animal and animal, and in some cases between plant and animal are discussed in Chapter 3. These are selected not only from the Platyhelminthes but from other phyla of the animal kingdom to emphasise the generality and universality of the phenomena.

Some aspects are only lightly touched on and one of these is the biochemistry of parasites, particularly the chemical reactions involved in nutrition, excretion, immunity and so on. These are aspects of the subject in which great advances have been made in the past two decades and fields in which there is a large and increasing number of research workers, but as this book is primarily for those who have not yet commenced a study of biochemistry this aspect of parasitism is not expanded.

I have a number of grateful acknowledgements to make. Some illustrations I have adapted from those of others and due acknowledgement of this is made in the various legends concerned. I am indebted to Dr James S. Scott for photographs of a polycercus of *Paricterotaenia* and to Dr Michael Burt for electron micrographs of cestode tegument, while Dr G. A. T. Targett has given me the benefit of his knowledge in discussing and making suggestions on that section of this book dealing with immunology. I thank Professor I. M. Sandeman of Trent University, Ontario, for the photograph from which the illustration on the front cover of this book is made. It is of rostellar hooks of the tapeworm, *Paricterotaenia burti*, described by him. There are many others, old students of mine, both in Ceylon and elsewhere, whose work under me has borne fruit, and to whom I am indebted. I make grateful acknowledgement to my wife, not only for her patience and encouragement but also for her help in typing the manuscript and compiling the index. Professor J. E. Webb is responsible for many suggestions and I thank him for the great care that he has taken in revising my manuscript.

St Andrews DAVID R. R. BURT

Contents

List of Figures

List of Plates

1 *Introduction*

All Nature seems at work.
S. T. COLERIDGE, *Work without Hope*

In the world of living things there are independent free-living forms as well as those manifestly dependent on others. Amongst the latter some live on, or within, the bodies of other organisms. Such dependent organisms are called parasites, and those which harbour them are called hosts. The meaning of the word parasite has changed from one who eats at another's table to a frequenter of the tables of the rich earning a living by flattery, and hence to the biological meaning of one organism completely dependent on another. However, there are various degrees of dependence. Scientific man with a mania for definitions and pigeon-holes has sorted out and defined different categories of association, but we find when we examine them closely that there is frequently an overlap between them or that a particular organism will fit as easily into one category as into another.

Parasites have no doubt been known to man from earliest times when he found them in himself or in animals of the chase. One would imagine that they would be accepted as belonging to the general scheme of things and just another manifestation of nature, but it is only within historical times that we can be certain of this. We have a learned account, by Baron von Oefele, of the parasites known to the Egyptians as long ago as the sixteenth century B.C. This account is mainly of the Ebers papyrus, but it refers to various other papyri including a veterinary one unearthed in 1889 at Kahun by Flinders Petrie. The parasites mentioned are all large or obvious, such as the tape-worm, the common round-worm, the whip-worm, the thread-worm and the guinea-worm, and the relation between them and disease is discussed. Although in the ancient Egyptian writings known to us there is no mention of the disease we now know as Schistosomiasis, a disease caused by small flukes inhabiting the blood system, the eggs of this parasite, which is still common in certain parts of Egypt today, have been identified in mummies that are 3,000 years old.

The taboos to which the Israelites were subject, as related in the Old Testament, doubtless had their origin in the observed effect of eating certain animals, such as the pig, which were then forbidden.

At a later date the Greeks were familiar with parasites and Hippocrates writes of them, and there is even mention of one in the comedy *The Horsemen* of Aristophanes where one of the characters invokes Jupiter and demands that a particular man have his mouth forced open and his tongue examined in the same manner as that of cooks in examining a pig before slaughter. Aristotle refers to the same parasite and the site of infestation, at the base of the tongue, where it can be readily seen in a living pig.

Parasitism is not a rare phenomenon, nor is it restricted to any one group. There is not a single phylum or major group of animals that does not contain forms that are parasitic, while there are even some where every member is a parasite. Parasitism is so widespread that it can be safely said that there are more parasitic animals than there are free-living ones, both in numbers of individuals and in kinds. It is doubtful if there exists a single species that does not harbour at least one parasite while in many the parasites that have been recorded may be counted by the dozen.

The literature on parasitism is extensive. It is not the aim of this book to give a summary of this, but rather to indicate the various kinds of association amongst animals and some possible ways in which parasitism may have had its origin. This of course leads naturally to a discussion of the evolution of parasitism in the different groups in which it is found with concomitant evolution of adaptations and variations in the life-cycle. We could equally express this in another way, that the evidence for evolution is nowhere more cogent than in a study of parasitism.

The close relationship between a parasite and its host is only one instance of the very special relationship that exists between an organism and its environment. The environment in this case is the body of the host, but it differs from an external environment in that the very presence of the parasite can alter it by inducing a reaction by the host. This reaction can be detrimental to the parasite in influencing its development or limiting its reproduction, or in preventing further infestation by similar or allied forms. An animal may have a natural immunity to a particular parasite, or this immunity may be acquired. On the other hand the host may tolerate only a limited number of parasites which is a condition known as *premunition*.

The intimacy of the special relationship between parasite and host is seen in those cases where the parasite is dependent on its host for the hormones that stimulate and control the parasite's own reproduction. This phenomenon is seen both in some internal and some external parasites, and whatever the origin may have been, the advantage to the parasite is an obvious one in that it reproduces at the same time as its host and dissemination of its progeny is thus ensured.

Another phenomenon that will be discussed is host-specificity, or the restriction of the parasite to a particular kind of host, or for that matter to a limited number of kinds of hosts. Some parasites are catholic in the variety of their hosts while others are never found other than in, or on, a particular host-species. This specificity is more common in some groups of parasites than in others and where it is found it is possible to trace the evolution of parasites *pari passu* with that of their hosts.

The modes of transference of the parasite from one host to another is seen in the many variations in life-cycles. The life-cycle of the parasite may be a direct one, not involving an intermediate host and infestation of the new host may be a matter of chance. On the other hand there may be one or more intermediate hosts with, or without, free-living stages in the life-cycle. Chance may play a part here too in the finding of a host or there may be a particular relationship between the main host and the intermediate ones that favours the transference of parasites. But there is compensation for these hazards in the high biotic potential of the parasite in its often prodigious powers of reproduction.

2 The Classification of Platyhelminthes

Carolus Linnaeus, *Systema Naturae*

The phylum Platyhelminthes comprises animals, commonly called flatworms, that show many differences in their form, behaviour and life-cycle, yet which present a number of common and fundamental characters that warrant grouping them together in one phylum. They are multicellular animals, or metazoa, that show a degree of development more advanced and more specialised than that of the coelenterates, with which group they share the common character of the absence of a coelom, or cavity within the body other than a gut or enteron. This cavity also has a single opening to the exterior. Unlike coelenterates, the Platyhelminthes are triploblastic, which implies that the tissues of their bodies are derived from three so-called germinal layers, or at least have three separate origins. The third layer, absent from the coelenterates as a distinct cellular layer, is the mesoderm, which in its fully developed state forms a packing of spongy, connective tissue, known as parenchyma, and it is in this that the various organ-systems of the body are embedded.

Flatworms are, for the most part, flattened and bilaterally symmetrical with one end of the body specialised as an anterior end: they lack a haemocoel, a circulatory system and a respiratory system. The excretory system is in the form of protonephridia, which are enlarged and hollow cells, called flame-cells, in the cavity of which there are cilia whose movement and flickering appearance give rise to this common name. These flame-cells function as excretory and osmoregulatory organs, and their excretions are carried through intracellular channels which unite and join longitudinal canals leading to the exterior. The reproductive system is independent of the excretory system, is frequently complex and usually hermaphrodite.

The animals called nemerteans, in the phylum Rhynchocoela, have much in common with Platyhelminthes, notably the absence of a coelom and the presence of parenchyma, but they differ in that there is a gut present which is frequently differentiated into distinct regions and always possesses two openings to the exterior, a mouth and an anus. They also possess a characteristic, eversible proboscis which lies above the gut in a tubular recess, called the rhynchocoel. As there is no other group of animals of a comparable grade of development one may define the Platyhelminthes briefly, and succinctly, as *acoelomate, bilateral, triploblastic metazoa in which an anus is lacking.*

It is not difficult to imagine the ancestral platyhelminth from which all modern forms have developed for it would be likely to possess many characters common to its descendants, and it is reasonable to assume that it would also have those features possessed by the more generalised or least specialised of its descendants. This ancestor had, most probably, the following form. It was flattened, covered with ciliated epithelium and with a mouth on the ventral surface towards the anterior end of the body. This end possessed cerebral ganglia and receptor organs and from the ganglia nerves extended through the length of the body. The mouth led into a simple sac-like intestine and between this gut and the ciliated epithelium there was a packing of mesenchyme cells. Muscle-fibres were differentiated in this mesodermal layer, and through it there passed dorso-ventral fibres, while at the periphery of it, just below the ciliated epithelium, were circular and longitudinal fibres. The excretory system consisted of flame-cells leading into two longitudinal canals which opened to the exterior. The reproductive system comprised paired ovaries and paired testes leading separately to the exterior, the oviducts of the former uniting and enlarged forming a spermotheca, and the vasa deferentia from the latter uniting to form an ejaculatory duct terminating in a copulatory organ. This organ was either a cirrus, the terminal eversible part of the duct, or a penis, the permanently evaginated terminal part of the duct with a thickened wall which thus formed a muscular conical projection.

The evolution of this ancestral form, giving rise to the many and varied forms living now, has been accompanied by an adaptive radiation similar to that seen in the evolution of other animals. The descendants, acquiring new features, both morphological and physiological, have spread into new biotopes. An idea of the variety and extent of these can be seen in a review of the classification of the phylum. As in all classifications, different workers vary in the relative importance which they place on the characters the animals show. The following classification is based on the nature of the larvae, the structure of the adult and on the nature of the life-cycle. On these grounds one recognises in the Platyhelminthes the following groups of worms which are sufficiently distinct to be called classes, although, as will be seen, some of them are more nearly related than are others.

- Turbellaria or Planarians.
- Temnocephalida.
- Monogenea or Monogenetic Trematodes.
- Aspidogastrea.
- Digenea, Trematoda, Digenetic Trematodes or Flukes.
- Didymozoonidea.
- Cestodaria or Monozoic Cestodes.
- Cestoda or Tapeworms.

CLASS TURBELLARIA

Turbellaria, or planarians, are for the most part free-living in the ocean, in fresh water or on land. Some of them, however, are phoretic, commensal or ectoparasitic and some few are endoparasitic. The outer covering of the body is a single layer of ciliated cells, the cilia being restricted to certain regions in some forms, even absent in others, or in a few cases they appear to have sunk into the body leaving the outer surface smooth and flat and without nuclei. A gut is present in the adult except in the Archoophora or acoelids. The main criterion used in earlier systems of classification was the form of this gut. There were thus grouped together those forms where the gut was present in a simple unbranched condition as in the Rhabdocoelida, or where there were three diverticula or limbs as in the Tricladida, or where the gut showed numerous radiating branches as in the Polycladida, or where it was an organ without a cavity and formed of a syncytial mass of cells as in Acoela. There were some forms that did not fit into this classification, such as those in which the gut had short diverticula. With these were associated features differing from those of the majority with simple sac-like intestine. These were separated into the order Allocoela. But there were many who did not regard this grouping as satisfactory because in some ways it ran counter to the arrangement of the female reproductive system, which character is now considered to be more fundamental than the nature of the gut.

In the more specialised Platyhelminthes there is a separation of the female gonad into a *germarium* in which ova are developed and a *vitellarium* in which yolk or vitelline cells are formed. Vitelline cells, as separate entities, come to be enclosed with an ovum within the egg-shell. In the Turbellaria, however, one finds a series of different stages in the evolution of the female reproductive system. There are forms in which the vitellarium is absent, those in which some cells in the gonad are nutritive cells and others are germ-cells or where, as is also found in tapeworms and flukes, there is a complete separation of the yolk-gland. There are other characters that show variation throughout the turbellarians such as the nature of the pharynx. This organ may be simple and unspecialised, or muscular when the form may be tubular, plicate or bulbous. The use of this character is, however, limited for whereas in forms with a much-branched gut the pharynx is always tubular (as it is in

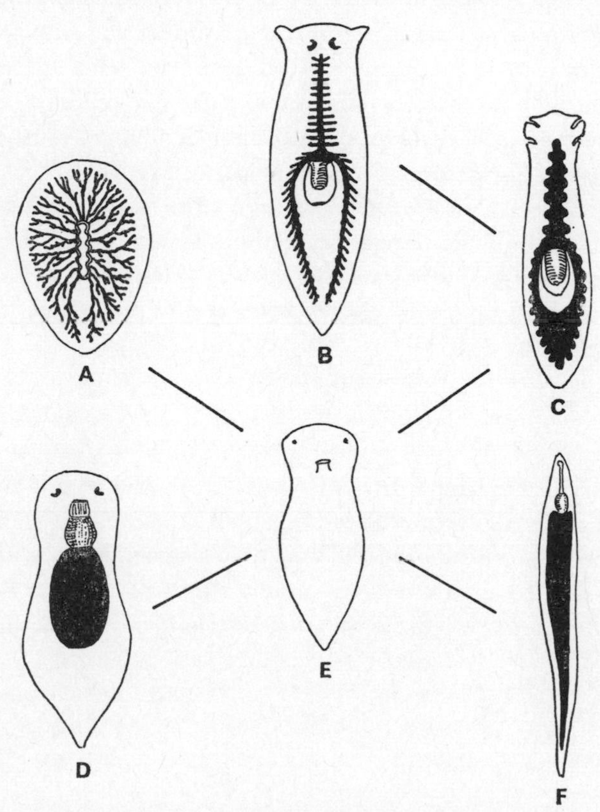

FIGURE 1 The orders in the class Turbellaria showing the characteristic form of the intestine in each order.

A, Polycladida: marine; intestine much branched; vitellaria are not separate; larva is free-swimming *Müller's Larva*; B, Tricladida: marine, fresh-water and terrestrial; intestine 3-branched; vitellarium is separate from germarium; C, Protricladida: marine and fresh-water; intestine incompletely divided into three branches; vitellarium and germarium separate; D, Eulecithophora: occupy every conceivable aquatic niche; intestine a simple sac; yolk cells are separate and not absorbed by developing ovum; E, Archoophora: fresh and brackish water; intestine lacking; vitellarium is not differentiated; F, Perilecithophora: fresh, stagnant and brackish water, and moist soil; intestine unbranched; developing ovum associated with yolk cells.

forms which show three main divisions of the gut) yet, in those in which the gut is unbranched, one finds every form of pharynx. Another character that varies is the type of segmentation of the egg during development and whether it develops into a free-swimming larva which undergoes a metamorphosis or develops directly into a miniature adult.

In the ancestral turbellarian there must have been present the latent potentiality to evolve into the great variety of forms seen today, but apparently this variation did not follow the same time-sequence in each group. Accordingly, one finds groups of animals showing dissimilar and similar characters, and the difficulty is to decide which characters are the most fundamental and base the classification on these. There is no palaeontological record to guide one in this choice so that whatever classification is adopted must either be an arbitrary one or an assessment of all the evidence available.

The most recent classification put forward takes as the most fundamental character the nature of the female reproductive system, and in accordance with this some of the older names of the various orders have been changed to terms which are descriptive of the ovarian condition. This classification contains the following orders.

Polycladida

Mostly large and flat marine forms with a much-branched gut and a tubular pharynx. The female reproductive system is simple without separate vitellarium; the eggs are endolecithal and their segmentation is spiral. This gives rise to a ciliated larva known as *Müller's Larva* (Fig. 2).

Examples: *Emprosthopharynx* Bock, *Euprosthiostomum* Bock and *Stylochus* Ehrnbg.

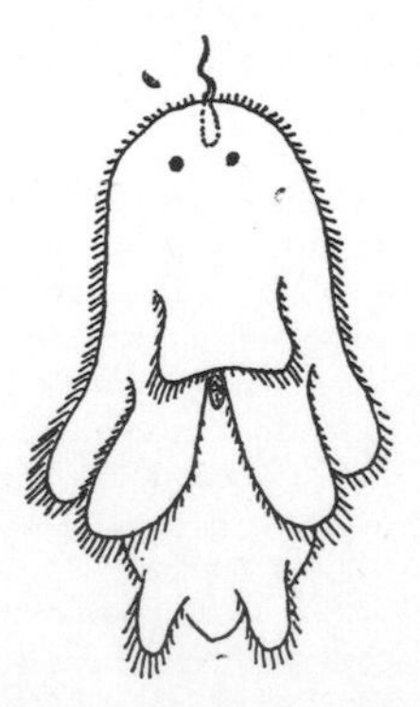

FIGURE 2 *Müller's Larva* of cotylean Polycladida.

Tricladida

Marine, fresh-water and terrestrial forms with a three-branched gut and tubular pharynx. They are mostly large and medium-sized forms. A vitellarium is distinct so that the eggs are ectolecithal and segmentation is mosaic, development being direct. The members of the Tricladida fall naturally into three sub-orders, Maricola, Paludicola and Terricola, names based on their habitat, and in these habitat and structure are in accord.

Examples: Maricola—*Bdelloura* Leidy (Fig. 3), *Micropharynx* Jägerskiöld; Paludicola—*Planaria* O. F. Müller, *Dendrocoelum* Oersted, *Kenkia* Hyman; Terricola—*Bipalium* Stimpson.

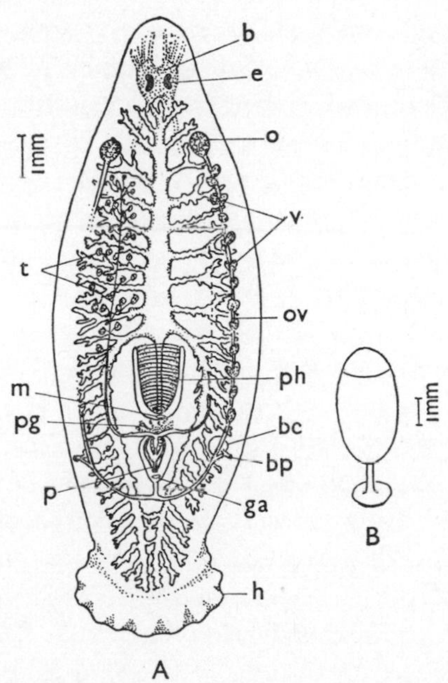

FIGURE 3 *Bdelloura candida* (Girard), a triclad turbellarian from the gills of *Limulus polyphemus* (L.)
A, entire animal; B, stalked egg; b, brain; bc, copulatory bursa; bp, bursal pore; e, eye; ga, genital aperture; h, haptor; m, mouth; o, ovary; ov, ovo-vitalline duct; p, penis; pg, penis-glands; ph, pharynx; t, testes; v, vitelline glands.

Protricladida

Fresh-water and marine forms with a pharynx similar to that of the triclads but with a simpler gut in that the posterior moiety of the gut is not completely separated into two branches. The vitellarium is distinct. In one sub-order development is exclusively parthenogenetic, while in the other knowledge of development is lacking. In other respects this is a primitive group for there is a statocyst similar to that found in coelenterates, but the statolith is accompanied by several accessory statoliths which is a condition found in the most primitive Turbellaria.

Examples: *Pseudomonocelis* Meixner (marine and fresh-water), *Ectocotyla* Hyman (phoretic on crabs and lamellibranchs) and *Peraclistus* Steinbock (phoretic on the crab *Hyas*).

Eulecithophora

This order comprises the old Rhabdocoela in which the gut is single and unbranched. The name is discarded as the character common to all members is the production of yolk cells which are not absorbed by the developing ovum. An egg-shell, operculate and sometimes pedunculate, is formed round the fertilised ovum, and development is ectolecithal and adiaphoretic. Apart from the simple, unbranched gut and the nature of the female reproductive system there is considerable variation. The pharynx is variable; it is bulbous, thick and muscular, functioning as a sucker in *Dalyellia*; in *Mesostoma* it is rosette-like and can be produced like a funnel; while in *Archimontresis* it is protrusible but remains tubular. The invasion of every conceivable aquatic niche, mud, sheltered and exposed beaches, interstitial water, fresh water, brackish water and the sea argues the antiquity of this group although there are few truly pelagic forms. The peculiar phoretic or commensal forms associated with fresh-water crustacea, the temnocephalids, are now considered as a separate class although obviously derived from dalyellians. There are truly parasitic forms, internal parasites of molluscs, echinoderms and crustaceans in addition to parasites of annelid and fish.

Examples: Sub-order Plagiostomatoida—*Archimontresis* Meixner, *Hypotrichina* Calandruccio, *Urastoma* Dörler; Sub-order Dalyellioida—*Dalyellia* Fleming, *Fecampia* Giard; Sub-order Mesostomatoida—*Mesostoma* Ehrb., *Phenocora* Ehrb., *Castrada* O. Schmidt.

Archoophora

This order, the least homogeneous in the class, comprises those forms in which a yolk-gland is not present as a differentiated organ. This character is shared with the polyclads but in the Archoophora a hollow gut is lacking as the digestive organ is formed of a syncytial mass of parenchyma representing both mesoderm and endoderm. This character would be diagnostic of the group did it not appear, at least temporarily, in other orders. Thus the endoderm and mesoderm have origin in a single blastomere as in polyclads; there are muscle-fibres in the epidermis as in some eulecithophorans; there is a statocyst similar to that of some protriclads; while the pharynx is simple in structure and usually a short ciliated canal like that which appears early in the development of polyclads. They are found in the sea and in fresh water and brackish water.

Examples: *Convoluta* Oersted, *Holfstenia* Bock, *Proporoplana* Reisinger.

Perilecithophora

In this very small order are placed those forms in which each developing ovum is surrounded by a group of vitelline cells. There are but two genera, restricted to the tropics but found in various habitats, viz. fresh water, brackish water, moist soil and even in the water collecting at the base of the leaves of pineapples. Although the initial stages of segmentation are reminiscent of spiral cleavage, yet the course of development is dependent on temperature. At lower temperatures the developing zygote comes to surround the vitelline cells, but in warmth the embryo develops in the middle of the mass of vitelline cells, at first absorbing the yolk through its surface but later through its mouth.

Examples: *Prorhynchus* M. Schultze, *Geocentrophora* de Man.

CLASS TEMNOCEPHALIDA

One of the main criteria used in the recognition of members of this class is their constant ecological association with other aquatic forms, mainly freshwater Crustacea. The common morphological characters are the possession of an adhesive organ, the presence of tentacles in the form of anterior or lateral processes, a simple sac-like gut, and the replacement of the external ciliated layer, common to free-living Platyhelminthes, with a naked syncytial layer.

The development of an adhesive organ and the loss of cilia are characters related to their attachment to other animals, while the simple gut is like that of a eulecithophoran turbellarian such as a dalyellioid. The posterior adhesive organ, or haptor, is like that of monogeneans, but monogeneans do not possess a syncytial layer. However, the one character peculiar to temnocephalids is the presence of tentacles. This outline of temnocephalid features can be filled in by noting the hermaphrodite reproductive system in which ovary and vitelline glands are separate, and in which the male and female ducts open in a common atrium on the ventral surface. There is no larval form, development being direct, the young temnocephalid hatching as a miniature adult from eggs which are attached, sometimes by a stalk, to the host-like partner.

It would appear that this group had its origin in the early adoption of a phoretic association with a crustacean. In the subsequent evolution this association has persisted although a few exceptions are to be met with in associations with molluscs and fresh-water tortoises.

Examples: *Temnocephala* E. Blanchard (Fig. 4), which lives in the branchial chamber of fresh-water Crustacea (Parastacidae) of New Zealand, Australia and South America; *Scutariella* Mrazek.

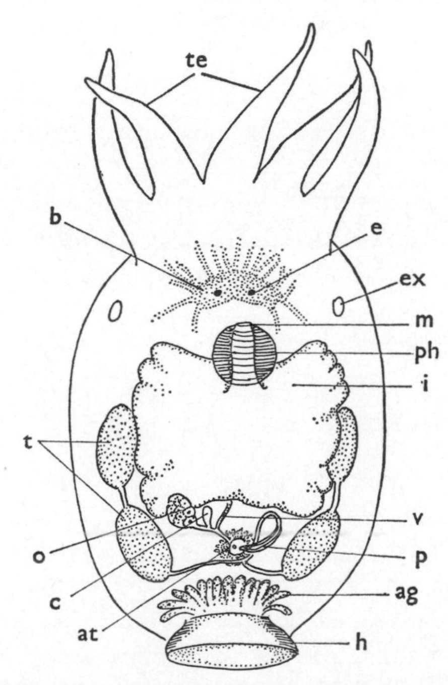

FIGURE 4 General anatomy of a temnocephalid (*Temnocephala* sp.) ag, adhesive glands; at, genital atrium and atrial glands; b, brain; c, copulatory bursa; e, eye; ex, excretory vesicle; h, posterior haptor; i, sac-shaped intestine; m, mouth; o, ovary; p, penis stylet; ph, globular pharynx; t, testes; te, tentacles; v, vitelline duct.

CLASS MONOGENEA

It is usual to regard Monogenea as ectoparasites living on a single host throughout the life-cycle, and this is mainly true although there are some which have become endoparasites.

Monogeneans possess organs of attachment in the form of a posterior haptor and two antero-lateral fields. The former shows a great variety in structure, some having a single sucker and others a number of them, the suckers being either naked or armed with hooks and spines, or in being modified as grasping clamps with a supporting skeletal mechanism (Fig. 5).

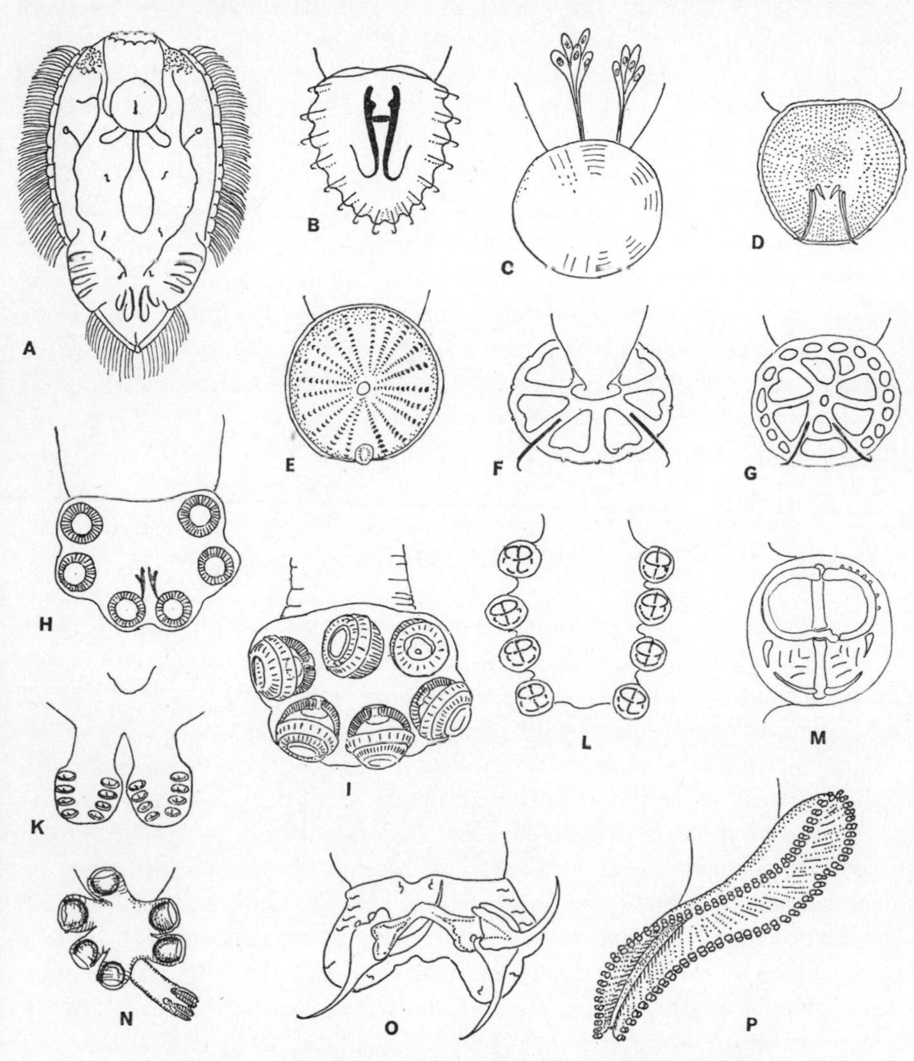

FIGURE 5 Evolution of the opisthaptor in monogeneans.
A, oncomiracidium of *Diclidophora denticulata* showing larval haptor (after Frankland); B, *Gyrodactylus elegans*; C, *Udonella caligorum* (after Johnston); D, *Entobdella hippoglossi*; E, *Acanthocotyle lobianchi* (after Monticelli); F, *Heterocotyle papillata* (after Doran); G, *Merizocotyle pugetensis* (after Kay); H, *Polystoma integerrimum*; I, *Oculotrema hippopotami*; K, *Diplozoon paradoxum*; L and M, *Diclidophora denticulata* and sclerites of clamp (after Frankland); N, *Rajoncocotyle emarginata* (after Price); O, *Tetraoncus monenteron* (adapted from Prost); P, *Microcotyle donavini* (after Sproston).

The two antero-lateral fields are glandular and produce a sticky secretion. It is obvious that the continued existence of these flatworms, clinging to the external surface of the body or to gills, often in swiftly moving hosts, has been possible by the evolution of efficient organs of attachment.

They are all hermaphrodite, the reproductive system comprising a single ovary with independent vitelline glands and usually a single testis, though in some cases there are numerous testes. The uterus usually leads into a genital atrium which also receives the vas deferens whose terminal region is modified as a copulatory organ, while the vagina, sometimes double, opens independently elsewhere. Large prostate glands may be associated with the vas deferens, while Mehlis's gland, concerned with the production of the egg-shell, opens into the ootype, the modified first region of the uterus. A feature of the monogeneans is a connection, in the form of a duct (the genito-intestinal canal) between the female reproductive system and the digestive system. This connection is usually with the oviduct but in the case of one group it is with the vitelline duct.

The body is covered with an acellular layer and immediately below this there are three layers of muscle-fibres: an external layer of circular fibres, an internal layer of longitudinal fibres and between them a layer of oblique fibres.

As a general rule monogeneans are oviparous and although in a few cases the young are hatched as miniature adults there is usually a free-living larva, generally ciliated, known as an *oncomiracidium* or *gyrodactyloid* larva which has a posterior haptor armed with small hooklets (Fig. 5A).

Various classifications of Monogenea have been put forward since the end of the eighteenth century, and although the names of the sub-divisions of the group have been changed and new sub-divisions added as new species were discovered, the main criterion used, until recently, in classification was the most obvious one, namely, the form of the posterior haptor. Since a natural classification should reflect the evolution and phylogenetic relationships of the organisms concerned, it has been maintained that the most fundamental character in Monogenea is the nature of the haptor, not of the adult but of the larva. Larvae are however known only in a handful of species, so that one has to fall back on the form of the adult haptor combined with the nature of the reproductive system. For the most part such a classification is in accord with the phylogenetic relationships of the hosts, or where there is disagreement it is in accord with their ecological relationships. Monogeneans can be grouped into two sub-classes depending on whether the haptor is a single structure or a multiple one, these are Monopisthocotylea and Polyopisthocotylea.

SUB-CLASS MONOPISTHOCOTYLEA

The posterior haptor is in the form of a single, well-developed disc, which may or may not be subdivided by radial septa, and which is armed with from one to three pairs of large hooks and from 12 to 16 marginal hooklets. The genito-intestinal canal is sometimes lacking. The Monopisthocotylea contains the following orders.

Capsaloidea

The number of testes is variable, one, two or more, but they lie posteriorly in respect of the ovary. They are oviparous.

Examples: *Entobdella* Blainville (Fig. 5D), *Capsala* Bosc, *Trochopus* Diesing, *Calicotyle* Diesing, *Merizocotyle* Cerfontaine (Fig. 5G), *Monocotyle* Taschenberg, *Dictyocotyle* Nybelin, *Dactylogyrus* Diesing, *Tetraoncus* Diesing (Fig. 5O), *Acolpenteron* Fischthal et Allison, *Tricotyle* Manter, *Tristoma* Cuvier, *Heterocotyle* Scott (Fig. 5F).

Gyrodactyloidea

The haptor is armed with 16 marginal hooks. A single testis is present lying in front of the ovary. They are viviparous, lacking both vitellaria and vagina.

Example: *Gyrodactylus* Monticelli (Fig. 5B).

Acanthocotyloidea

The haptor is armed with 16 hooks. A vagina is present.

Example: *Acanthocotyle* Monticelli (Fig. 5E).

Protogyrodactyloidea

In addition to marginal hooks there are two pairs of large hooks with supporting bars. There is a single testis. The vitelline glands are restricted to two or three groups and a vagina is absent.

Example: *Protogyrodactylus* Johnston et Tiegs.

Udonelloidea

Cylindrical in shape with an unbranched sac-like gut and a powerful but unarmed sucker-like opisthaptor. Other characters, reminiscent of temnocephalids, are the absence of a vagina and direct development, the young hatching as an immature adult from a stalked egg which is attached to the host. Udonellids live on crustaceans, but unlike temnocephalids they are associated with marine and not fresh-water hosts.

Example: *Udonella* Johnston (Fig. 5C).

SUB-CLASS POLYOPISTHOCOTYLEA

The haptor comprises a number of suckerlets or clamps. A genito-intestinal canal is present. The Polyopisthocotylea contains the following orders.

Chimaericoloidea

The haptor of the *oncomiracidium* is armed with 16 hooks. The opisthaptor of the adult is borne at the end of a long posterior stalk or peduncle and comprises four pairs of clamps.

Example: *Chimaericola* Brinkmann.

Diclidophoroidea

Apart from the chimaericoloids this order is the only one possessing clamps. They are generally paired, four being the usual number though sometimes more numerous and arranged asymmetrically. They possess two small suckers within the buccal tube. The haptor of the *oncomiracidium* is armed with ten marginal hooklets and two or four median hooks.

Examples: *Diclidophora* Diesing (Fig. 5A, L, M), *Cyclocotyla* Otto, *Anthocotyle* v. Beneden et Hesse, *Diplozoon* Nordmann (Fig. 5K), *Mazocraës* Hermann, *Kuhnia* Sproston (syn. *Octostoma* Kuhn), *Microcotyle* v. Beneden et Hesse (Fig. 5P).

Polystomatoidea

At the anterior end the mouth lies within a terminal depression and at the posterior end the opisthaptor comprises one pair or three pairs of unarmed suckers and usually a pair of large hooks. A single testis is present but it is frequently follicular and appears as numerous testes. There are 16 hooklets on the larval haptor. They are ectoparasites and endoparasites of amphibia and reptiles and one species is an ectoparasite of the eye of a hippopotamus.

Examples: *Polystoma* Zeder (Fig. 5H), *Oculotrema* Stunkard (Fig. 5I), *Sphyranura* Wright.

Diclybothrioidea

The mouth lies within a sucker at the anterior end. The opisthaptor may be borne on a stalk but it comprises three pairs of suckers differing from those of the Polystomatoidea in that each sucker is supported by a semicircular sclerite one end of which is pointed and protrudes through the surface. In addition, extending beyond the haptor, there is sometimes a terminal process which is armed with a pair of suckers and a pair of hooks. Testes are numerous. The larval haptor is armed with ten hooklets and two pairs of median hooks.

Examples: *Rajoncocotyle* Cerfontaine (Fig. 5N), *Diclybothrium* Leuckart.

CLASS ASPIDOGASTREA

This small group of parasitic worms is distinguished by the possession of an elongate adhesive disc, like the sole of a shoe, which covers the greater part of its ventral surface and which is formed of a number of alveoli. In most aspidogastreans the disc has the appearance of a long sucker subdivided by transverse septa into separate alveoli, or showing further subdivisions by longitudinal septa, although in one genus (*Stichocotyle* Cunningham) (Fig. 7) the adhesive organ is in the form of a single row of small and separate suckers. A characteristic feature of many forms is the possession of sense organs known as *marginal organs*. Each lies at the lateral margin of a transverse

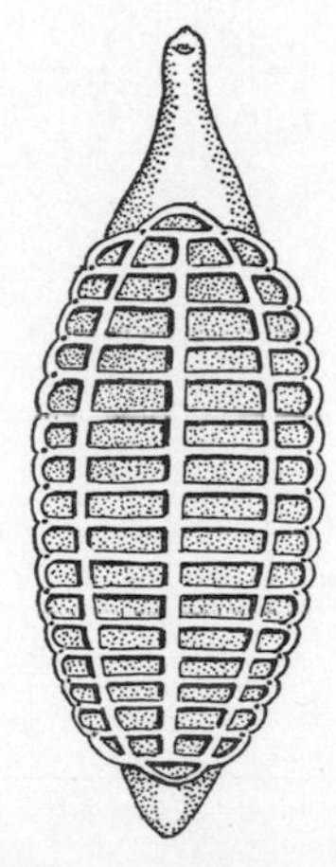

FIGURE 6 *Aspidogaster conchiola* v. Baer from the pericardium of species of the mussels *Anodonta* Lamarck and *Unio* Retzius. Almost the entire ventral surface is covered with a large compound adhesive organ formed of numerous alveoli. Marginal organs are present at the lateral margins of the transverse septa. (2·5 to 3 mm long.)

septum and leads through a duct to a mass of neuro-secretory cells. The mouth is at the base of an oral funnel, modified in some forms as an oral sucker, and is followed by a muscular pharynx and an unbranched sac-like intestine. The excretory organs open through a single aperture posteriorly, although in the larva there are two separate openings. The larva is a *cotylocidium*, so called on account of its possession of a sucker which later develops into the adhesive disc. In its most primitive form the cotylocidium is free-swimming, possessing two discontinuous bands of cilia, but it can also move leech-like, using alternately the oral and the ventral sucker; in other forms the cilia are wanting.

There is a wide range of hosts in which they occupy a variety of sites; in the intestine of rays and the gall-bladder of chimaeras, in the intestine of fresh-water tortoises and teleosteans, in the branchial cavity and pericardium of bivalves and among the ctenidia of prosobranch gasteropods. This phenomenon of parasitism within hosts belonging to two different phyla, Mollusca and Vertebrata, is reminiscent of that obtaining in Digenea, but in this case some species, such as *Aspidogaster conchiola* v. Baer (Fig. 6) and *Cotylaspis insignis* Leidy, attain sexual maturity in a mollusc and are not found

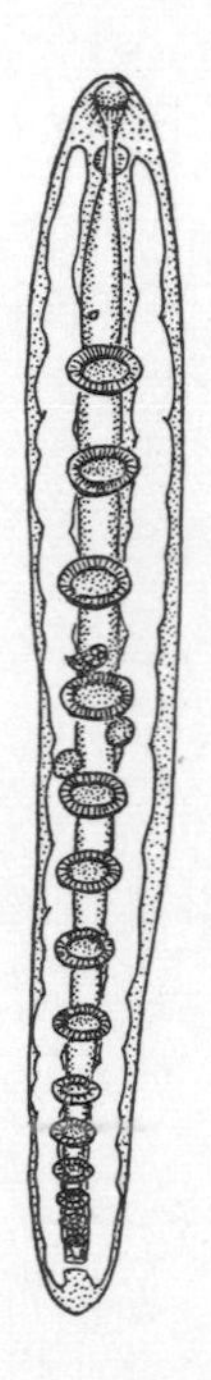

FIGURE 7 *Stichocotyle nephropis* Cunningham from the bile duct of *Raja clavata* L. A young individual in which the complete series (24–30) of ventral suckers is not yet developed. Marginal organs are absent from this genus. Length, when extended, may measure 115 mm.

in vertebrates. It could be argued that aspidogastreans were originally parasites of molluscs and that some of them later came to occupy niches in vertebrates, on the other hand parasitism in a mollusc might be regarded as a neotenic development.

The variation in the class is no more than that seen in most families and, accordingly, aspidogastreans are grouped in a single family Aspidogastridae in which two sub-families are recognised, each characterised by common morphological as well as ecological characters. The Aspidogastrinae contains forms parasitic in fresh-water molluscs, bony fishes and tortoises, while the

Stichocotylinae comprises forms found in marine cartilaginous fishes such as rays and chimaeras.

Examples: Aspidogastrinae—*Aspidogaster* v. Baer (Fig. 6), *Cotylaspis* Leidy. Stichocotylinae—*Stichocotyle* Cunningham (Fig. 7), *Macraspis* Olsson (Fig. 17).

CLASS DIGENEA

It is just a century and a half since the first scheme of classification of the helminths was put forward by Rudolphi, who invented the term Trematoda to cover unsegmented worms known before then as sucking worms. The name referred to the obvious character of the presence of suckers, which to the unaided eye looked like holes, the equivalent of which, in Latin, is *tremata*. The wide assemblage of worms included under this term is now restricted, the monogenetic trematodes being separated as the class Monogenea, and other worms with common, but fundamental, differences being likewise removed, leaving the great bulk of digenetic forms to constitute the now restricted Trematoda. It is a moot point whether the historical but inaccurate term Trematoda should be discarded in favour of Digenea. The latter term, implying the common character of the indirect or 'digenetic' mode of reproduction was used originally to designate one order, but it is now applied to a more restricted group and raised to the rank of class, one of the main divisions of the phylum. As the digenetic nature of the life-cycle is such a fundamental character this has altered the balance of opinion in favour of the use of the term Digenea.

There are two outstanding differences between the Digenea and the Monogenea, the first is seen in the mode of development and the second in the nature of the life-cycle. In development there are two constant larval forms. The first is called a *miracidium*, usually ciliated, which in the intermediate host undergoes a peculiar form of repeated asexual reproduction, variously described as parthenogenesis, metagenesis, paedogenesis or polyembryony, to give rise to the second form, the *cercaria*. In the life-cycle there is an alternation between two hosts, the first, or primary host, in which the adult sexual individuals are found, being a vertebrate, and the second, in which asexual reproduction takes place, being an invertebrate and almost invariably a mollusc. In addition there may be other and succeeding intermediate hosts, as many as three having been recorded, but it is the alternation between two hosts, the vertebrate and the mollusc, that is the fundamental character of the Digenea.

Most Digenea are oval, flattened creatures with specialised organs of attachment in the form of suckers or hooks, or of both. The body is covered with a cuticle or tegument, devoid of cilia, which rests directly on underlying mesenchyme cells. The digestive system usually shows an intestine which is bifurcated, the two branches ending blindly, and in nearly all there is but the single opening of the digestive system, the mouth.

They are mostly hermaphrodite with two testes, a single ovary and a common genital atrium into which open the vas deferens and the uterus, the latter also functioning as a vagina.

The majority are endoparasites living within various cavities of the body, including the blood-vessels, but a few live within cysts. Some few approach an ectoparasitic existence in the gill-chamber of fishes (*Gonocera* Manter, *Bathycotyle* Darr, *Syncoelium* Looss and *Otiotrema* Setti) while *Paronatrema* Dollfus is actually an ectoparasite on the devil-fish *Ceratoptera* Müller et Henle.

In the same manner as in the Monogenea one looks for fundamental characters in the larval forms, particularly the *miracidium* and the *cercaria*. Miracidia show differences in the number and arrangement of the excretory organs, the protonephridia, as well as in the presence of penetration glands and eyes. Cercariae can likewise be characterised by their nephridial arrangement, the nature of the excretory bladder, whether present primarily or developed secondarily, the presence of penetration glands or stylets, the presence of eyes and the easily recognised character of the form of the tail. The tail may be reduced or absent, and if present may be simple or forked, naked or provided with fins or cilia, and so on. Using these characters one can group species into orders and groups of families, and the arrangement so obtained is a natural one so far as our knowledge of the life-cycles goes.

Older classifications took cognisance of the obvious features of the organs of attachment in the adult. Thus the monostomes were those with a single sucker; the distomes had two suckers, an oral one and a ventral one; the amphistomes had two suckers, an anterior one, the oral sucker and another, usually a large one, at the posterior end of the body; the echinostomes were variously armed, usually with a circlet of hooks near the mouth; while in the schistosomes the sexes were separate.

In the most modern classification an attempt is made to use larval characters, but as a knowledge of the life-cycles of very many species is lacking there is a large residuum of unallocated families and accordingly a complete classification must await further discoveries. In this classification there are recognised two sub-classes or super-orders, the Anepitheliocystida and the Epitheliocystida.

SUB-CLASS ANEPITHELIOCYSTIDA

The primitive excretory vesicle of the cercaria persists in the adult state, the excretory pores open to the exterior in the tail and the cercaria is not armed with a stylet. The Anepitheliocystida contains the following orders.

Strigeatida

The *cercaria* generally possesses a forked tail and is known as a *furcocercaria*; it also possesses penetration glands and one or two pairs of protonephridia.

Examples: *Schistosoma* Weinland (Fig. 22), *Alaria* Schrank (Fig. 24), *Strigea* Abildgaard, *Diplostomum* v. Nordmann, *Cyclocoelum* Brandes, *Leucochloridium* Carus.

Echinostomatida

The *cercaria* possesses a simple tail and many cystogenous glands and the *miracidium* has a single pair of protonephridia. Cercariae develop within *rediae*.

Examples: *Echinostoma* Rudolphi, *Campula* Cobbold, *Himasthla* Dietz, *Parorchis* Nicoll (Fig. 19), *Philophthalmus* Looss, *Paramphistomum* Fischoeder, *Fasciola* L.

Renicolida

The *cercaria* is a bright rose colour, hence the name *Rhodometopa* applied to it; it has a Y-shaped excretory bladder with numerous lateral diverticula which is a

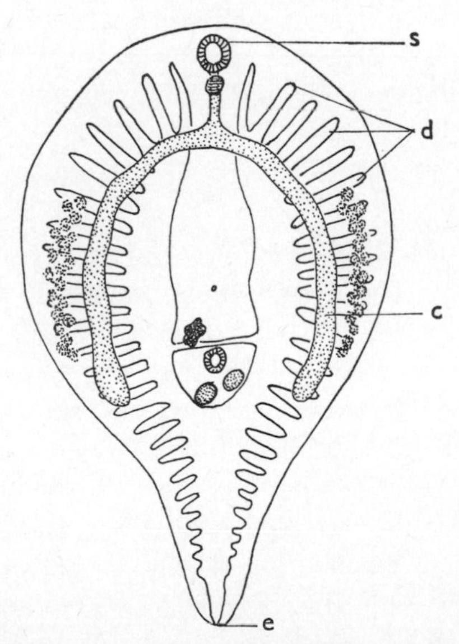

FIGURE 8 *Renicola* Cohn. Semi-diagrammatic representation showing the characteristic lateral diverticula (d) of the Y-shaped excretory bladder; c, intestinal caecum; e, excretory aperture; s, oral sucker.

character persisting in the adult. They are parasites in the kidneys of birds and have three hosts in the life-cycle, another being a marine mollusc *Turitella* in which the cercariae develop within sporocysts. There is a single genus *Renicola* Cohn (Fig. 8).

SUB-CLASS EPITHELIOCYSTIDA

The wall of the primary excretory vesicle is replaced by a thick layer of mesodermal cells. In these the cercariae have a simple and small tail, or none at all, and when a tail is present it rarely contains excretory ducts while the miracidium has but one pair of protonephridia. The Epitheliocystida contains the following orders.

Plagiorchidida

The *cercaria* is usually armed with a stylet and called a *xiphidiocercaria*. It generally encysts in an arthropod, there being three hosts in the life-cycle. The excretory vessels of the cercaria do not open to the exterior in the tail.

Examples: *Dicrocoelium* Dujardin (Fig. 23), *Haplometra* Looss, *Paradistomum* Kossack.

Opisthorchidida

The *cercaria* is never armed with a stylet while its excretory vessels always open in its tail.

Examples: *Heterophyes* Cobbold, *Cryptocotyle* Lühe, *Opisthorchis* Blanchard, *Clonorchis* Looss (Fig. 21), *Gonocera* Manter, *Bathycotyle* Darr, *Halipegus* Looss.

The following genera are referred to in the text but as their cercariae are not known they are not included in the foregoing classification: *Otiotrema* Setti, *Syncoelium* Looss, *Paronatrema* Dollfus.

CLASS DIDYMOZOONIDEA

Didymozoonids show so many distinctive features that they cannot be included with either the Monogenea or the Digenea. The larva (Fig. 9), in those life-cycles where it is known, lacks cilia and is armed at its anterior end with a circlet of alternating longer and shorter spines. The larvae, without the intermediation of another host, penetrate the bodies of the definitive host and there they become encysted, often in pairs. The adults have the curious form of a narrow thread-like anterior region arising from the end or the side of the much swollen remainder of the body. Fundamentally, they are hermaphrodites but they show suppression of the reproductive glands of one or the other sex which gives rise, in some cases, to dioecious forms, a result very different from that obtaining in the digenean schistosomes where sex is determined genetically.

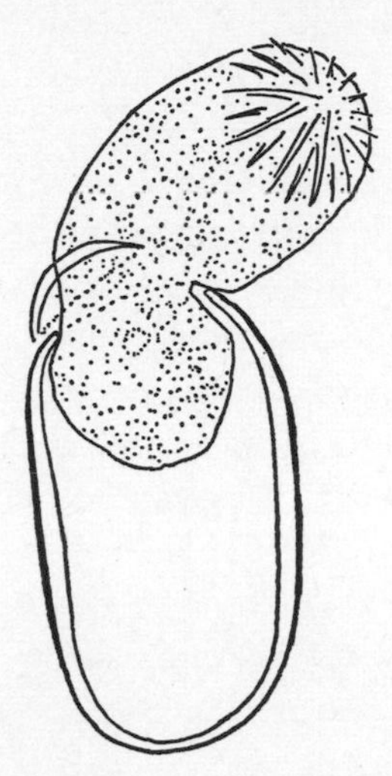

FIGURE 9 *Didymozoon faciale* Baylis. Egg, with embryo escaping under pressure of cover-glass. (Adapted from Baylis.)

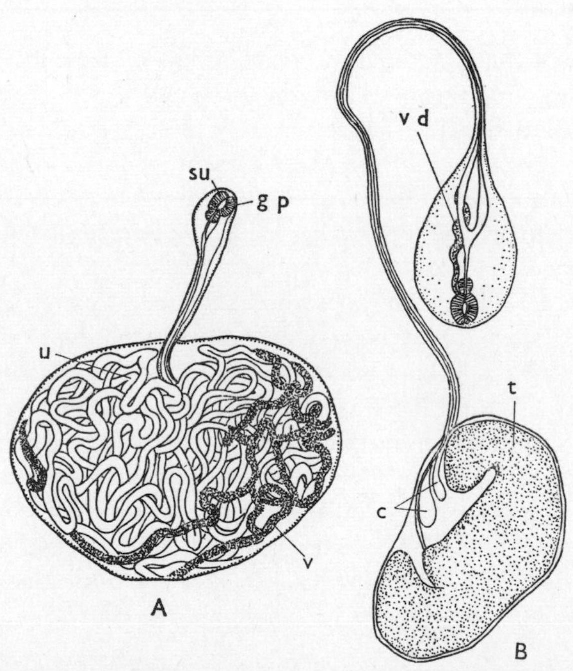

FIGURE 10 *Wedlia orientalis* Yamaguti from the gills of the tunny *Thunnus thynnus* (L.). A, female and B, male; c, intestinal caeca; gp, genital aperture; su, oral sucker; t, testis; u, uterus; v, vitelline gland; vd, vas deferens (after Yamaguti).

Didymozoonids are found, with two exceptions, in marine fishes, encysted in various organs such as the gills, the head, in the buccal cavity and within muscles.

As the larva differs from those of other Platyhelminthes and as the development is a direct one, those forms must be considered as belonging to a separate and distinct class, although one sees an evolutionary convergence in the absence of cilia as occurs in the digenean *Halipegus* Looss and the presence of spines on the larva which is also reminiscent of that genus.

The genera of the Didymozoonidea are few and can be grouped in a single family, the Didymozoonidae and they can be defined according to whether the separation of the sexes is complete as it is in *Wedlia* Cobbold (Fig. 10), or not complete as in the hermaphrodite genus *Didymozoon* Taschenberg (Fig. 9). Other characters which distinguish the different genera are minor ones such as the presence of one or two testes, a single or a double ovary and whether the ovary is branched or not.

CLASS CESTODARIA

The cestodarians comprise another small group of parasitic forms found mainly in the intestine or body-cavity of fishes and which stands apart from other Platyhelminthes. They were formerly thought to be related to tapeworms and were classed with them as unsegmented, neotenic forms as indeed was the case of some worms included in the group, but a fuller knowledge of their larvae, anatomy and life-cycles made it abundantly clear that they constitute a very distinct and separate group.

Cestodarians are Platyhelminthes, unsegmented, lacking a gut, hermaphrodite, with a single set of reproductive organs the ducts of which open separately. A scolex is lacking but there is a powerful muscular organ at one end of the body. Within the class there are two groups of animals having marked differences in structure but linked by the further common factor of having the same larval form, a *lycophore*, a larva with five pairs of hooks grouped at one end of the body and large glands opening at the other. These groups, the less-known Amphilinoidea, parasitic in the coelom of some fishes and chelonians, and the Gyrocotyloidea, parasites in the intestine of chimaeras, are usually considered as separate orders within the class (Fig. 11).

Amphilinoidea

The surface of the body lacks spines though it is frequently rugose due to the presence of powerful peripheral muscles while the muscular organ at the anterior end is in the form of a proboscis. The apertures of the cirrus and vagina open independently but close together at the posterior end of the body. The uterus, with origin in the posteriorly placed ovary is in the form of the letter N, showing three limbs in its arrangement throughout the body before opening at the anterior end close to the muscular proboscis. The testes extend through the greater part of the medulla in two fields, while the vitellaria, likewise extensive, lie lateral to the testes in two narrow peripheral bands.

Example: *Amphilina* Wagener.

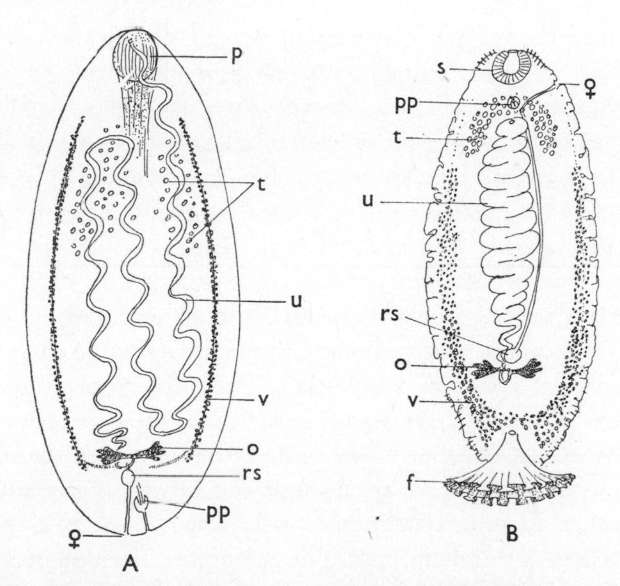

FIGURE 11 Cestodaria. Diagrammatic representation of A, an amphilinoid and B, a gyrocotyloid, showing the salient differences.
f, funnel; o, ovary; p, proboscis; pp, penis papilla; rs, receptaculum seminis; s, sucker; t, testes; u, uterus; v, vitelline glands; o, ovary; ♀, vaginal aperture.

Gyrocotyloidea

The surface of the body, often wrinkled transversely, is armed with spines which are not simple cuticular spines as found in flukes, for they show a concentric structure and they are movable, lying deeply in the cortex and operated by muscles which are associated with the muscular layer which separates the cortex from the medulla (Fig. 27). The muscular organ at the anterior end of the body is in the form of a powerful sucker while at the posterior end there is a funnel-like organ often described as a rosette on account of its fimbriated and much-frilled margin, the funnel opening dorsally through a small pore. The testes occupy two lateral fields anteriorly while the ovary, lying close to a large receptaculum seminis, is situated in the posterior moiety. The uterus is a conspicuous, much-convoluted organ occupying the centre of the medulla and it opens anteriorly as do the vas deferens and vagina. So far as is known the life-cycle does not involve an intermediate host.

Examples: *Gyrocotyle* Diesing (Plate 2), *Gyrocotyloides* Fuhrmann.

CLASS CESTODA

Cestodes, or tapeworms, are always endoparasitic and in the sexually reproducing or adult form are found, with one exception, in vertebrates, lying generally in the gut. There is considerable variation in the shape of the body, but the most usual arrangement is that of a segmented, flattened, elongate region, called the *strobila*, which is separated from the organ of attachment, called the *scolex*, by a growing *neck* where the segments, usually referred to as *proglottides*, are formed. This is the most common appearance, exceptions being seen in some worms where the body is rounded and not flattened, in others where an external segmentation is not apparent, and in others where segmentation takes place within already formed proglottides and not only in the region between scolex and strobila. They are usually hermaphrodite with a single set of reproductive organs in each proglottis, although exceptions are found in dioecious forms or where there is duplication of the reproductive organs of one or both sexes within a single proglottis. There are, however, certain characters that are universal within the class and, accordingly, diagnostic. These are the endoparasitic condition, the absence of a gut at every stage of the life-cycle and the nature of the larva.

The larva, a six-hooked body known as a *hexacanth* or *oncosphere*, is enclosed within an embryophore which in some cestodes is a ciliated covering and in others a non-ciliated protective coat. In the former case the hexacanth with its ciliated covering is known as a coracidium (Fig. 30h). There is a semantic difficulty here, for the coracidium, on hatching, is a free-swimming body and can accordingly be termed a larva. However, that term can hardly be applied to the corresponding and homologous, non-ciliated embryophore with its contained hexacanth, though in each case there is eventually liberated the *hexacanth* which is indeed a larva that actively penetrates the tissues of the intermediate host (Fig. 29e and f).

As a general rule the life-cycle involves one or more intermediate hosts which may be vertebrates or invertebrates and, in addition, there may be paratenic hosts and hosts in transit.

In addition to the nature of the embryophore one uses as criteria in classification various differences in adult morphology, particularly the nature of the attaching organs of the scolex and the arrangement of the reproductive organs. Seven orders can be recognised.

Haplobothriidea

This order contains a single genus *Haplobothrium* Cooper, parasitic in the American bow-fin *Amia calva* L., the only survivor of a formerly extensive group of fishes. The scolex lacks suckers or bothria but possesses four retractile proboscides which are armed with spines at their base. The primary strobila forms a series of secondary strobilae throughout its length and these break away, each with a pseudoscolex which, however, lacks proboscides. The larva is a ciliated *coracidium* and there are two intermediate hosts, a cyclops in which the larva develops into a *procercoid*, and a fish in which a *plerocercoid* is formed. As the arrangement of the reproductive system is similar to that seen in the Pseudophyllidea, the Haplobothriidea combines characters seen in that order and in the Tetrarhynchidea in which there are armed proboscides.

One species is known, *Haplobothrium globuliforme* Cooper (Fig. 12A).

Pseudophyllidea

The scolex is rarely armed, the organ of attachment comprising dorsal and ventral bothria which may be reduced to two grooves with muscular walls. As a rule the testes are numerous and scattered throughout the medulla, while the ovary is compact and the vitellaria lie laterally in the cortex. The uterus is usually tubular and sinuous with an opening to the exterior. The larva is a *coracidium* and there are two intermediate hosts in which the developmental forms, *procercoid* and *plerocercoid*, are found. Exceptionally, as in the family Caryophyllaeidae, there may be one intermediate host, or more rarely the development may be direct. They are parasitic in fishes but are also found to a limited extent in reptiles, birds and mammals which feed on fishes, the hosts of the plerocercoid stages.

Examples: *Diphyllobothrium* Cobbold (Figs. 12D, 30), *Bothridium* Blainville (Fig. 12E), *Duthiersia* Perrier (Fig. 12G), *Archigetes* Leuckart, *Ligula* Bloch, *Diplocotyle* Krabbe (Fig. 12I), *Diplogonoporus* Lönnberg.

Tetrarhynchidea

This order is easily recognised by the organs of attachment which are four proboscides armed with spines, much more highly specialised structures than those of the Haplobothriidea in being operated by an elaborate mechanism

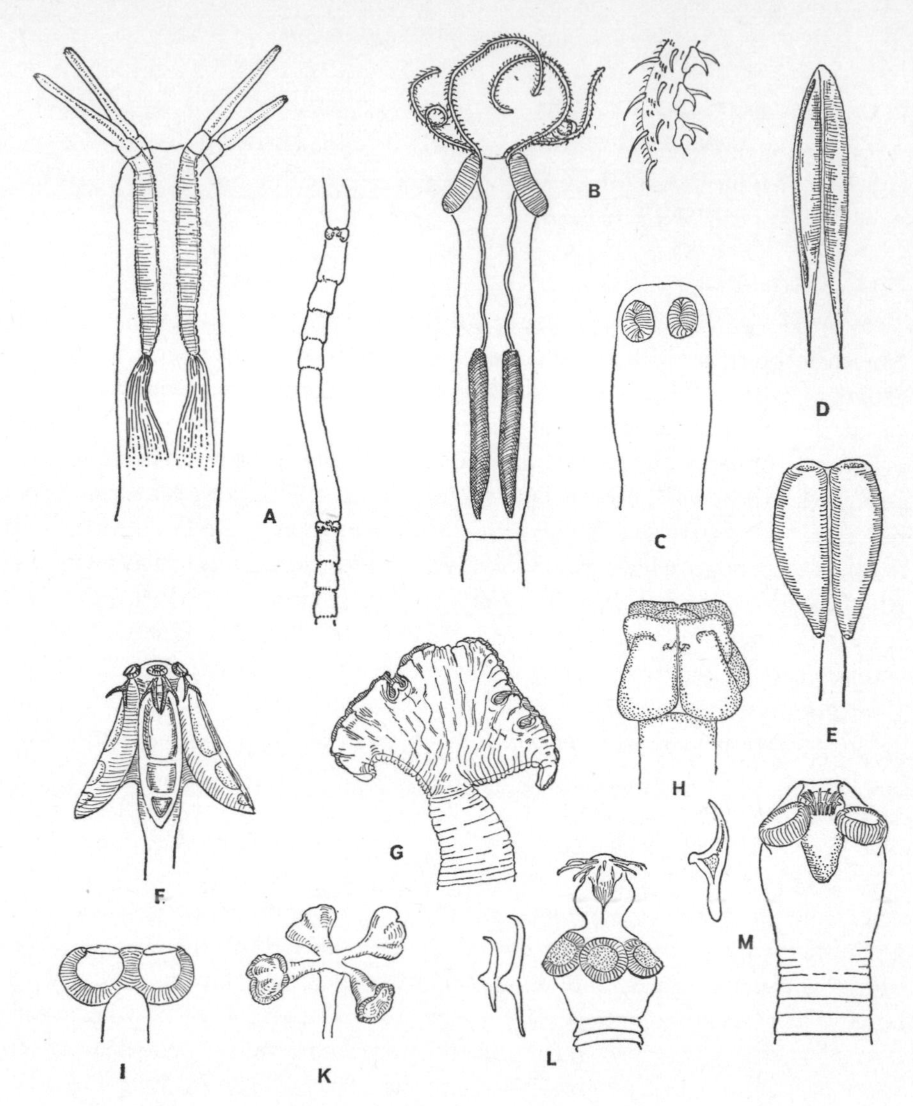

FIGURE 12 Scolices of cestodes showing some of the variations. A, *Haplobothrium globuliforme* Cooper, scolex and part of strobila which is dividing to form separate strobilae each with a pseudoscolex, from the North American bowfin, *Amia calva*; B, *Grillotia* sp. scolex and detail of proboscis, from the thorn-back ray, *Raja clavata*; C, *Ophiotaenia phillipsi* Burt, from the green pit-viper, *Trimeresurus trigonocephalus*; D, *Diphyllobothrium latum*, from man; E, *Bothridium pithonis* Blainville, from the Indian python, *Python molurus*; F, *Acanthobothrium* sp. from a shark; G, *Duthiersia fimbriata* (Diesing), from the tallagoya, *Varanus benghalensis*; H, *Tetrabothrius* sp. from the frigate bird, *Fregata ariel*; I, *Diplocotyle olrikii* Krabbe, from the amphipod, *Gammarus*; K, *Anthobothrium* sp. from the viviparous shark, *Scoliodon* sp.; L, *Parvitaenia ardeolae* Burt, scolex and rostellar hooks, from the Indian pond heron *Ardeola grayi*; M, *Onderstepoortia lobipluviae* (Burt), syn. *Choanotaenia l.*, scolex and rostellar hook, from the yellow-wattled lapwing, *Lobipluvia malabarica*.

in their extension and retraction. In addition, there are two or four bothria. The vitelline glands extend through the greater part of the cortex. They are parasites of selachians and, so far as is known, the larva is a *coracidium* and there are two intermediate hosts.

Example: *Grillotia* Guiart (Fig. 12B).

Tetraphyllidea

There are grouped in this order those parasites of elasmobranch fishes (rays and sharks) in which the hold-fast organ is in the form of four outgrowths of variable form, ear-like, trumpet-like or leaf-like, in which the margins may be expanded and fimbriated. In addition true suckers may be present and armed with single or double hooks. So far as is known the larva is a *coracidium* and there are two intermediate hosts.

Examples: *Acanthobothrium* v. Beneden (Fig. 12F), *Anthobothrium* v. Beneden (Fig. 12K).

Tetrabothriidea

In this small group the general arrangement of the reproductive system is that of a cyclophyllidean, except that the compact vitelline gland lies in the anterior region of the proglottis and not posterior to the ovary as in that order. The genital atrium is lateral and very muscular, while the uterus is sac-shaped. The hold-fast is variable: there may be four suckers with or without auricular appendages, or the suckers may be reduced and even absent, the reduction being usually accompanied by a corresponding development of the apical region of the scolex. Development is not known, but the intermediate host being a fish is the ecological factor governing the presence of these cestodes in Cetacea, Pinnipedia and marine birds.

Example: *Tetrabothrius* Rudolphi (Fig. 12H).

Ichthyotaeniidea

The tapeworms grouped in this order have four suckers similar to those of the Cyclophyllidea and there may even be an apical sucker, but the arrangement of the reproductive system is on the same plan as that seen in the Tetraphyllidea. The larval form is a *coracidium* and two intermediate hosts are known, but in some cases there is suppression of the second intermediate host. In the latter case the young fish, while still a ciliary feeder, becomes infested with a procercoid present in a copepod. They parasitise fresh-water fishes, amphibians and reptiles.

Example: *Ophiotaenia* La Rue (Fig. 12C).

Cyclophyllidea

There are few exceptions to be noted in the definition of this order which comprises the great number of tapeworms parasitising birds and mammals. There are four suckers in the scolex, and a terminal muscular rostellum, if present, may be unarmed or armed with hooks. The vitelline gland is compact and lies behind the ovary, and except in one family Mesocestoididae, the genital pores are lateral. Uterine pores are wanting, eggs being liberated from the host in proglottides which are shed and which later split open or otherwise disintegrate. The larva is not ciliated and there is generally a single intermediate host, in which, if it is a vertebrate, the hexacanth develops into a form of bladderworm, or, if it is an invertebrate, into a *cysticercoid* which is a solid form. In a very few instances an intermediate host can be dispensed with and then the development is direct.

It is a curious fact that in most classes of animals there is one order, much larger than the others, in which evolution has apparently run riot, as for instance the Coleoptera (beetles) amongst insects, the Passeriformes (perching birds) amongst birds or the Rodentia amongst mammals. The class Cestoda is no exception, for the Cyclophyllidea is by far the largest and, in virtue of cyclophyllideans being parasites of birds and mammals, is the latest to evolve. The principal variations seen in the order are the following.

The uterus may be sac-shaped, spherical or a network, or it may be replaced by, or resolved into uterine capsules, or it may be enclosed in one or more paruterine organs. The genital apertures may open on one side of the body or the other, or alternate from one side to the other, regularly or irregularly, or the reproductive system may be duplicated in each segment in which case the genital apertures open on both sides. This duplication of the reproductive system does not appear to be a fundamental character for it is found in worms which are obviously related to others in which the reproductive organs are single in each segment, both being however related by common morphological characters as well as by the nature of the host. Some depart from the general rule that cestodes are hermaphrodite and these also show the character of the absence of a vaginal aperture, a character shared with related hermaphrodite forms. In dioecious tapeworms all the segments of one worm are male and all those of another are female, pairs of worms being found together in the same host. A variation in the arrangement of the reproductive organs is seen in *Gynandrotaenia* Fuhrmann, a genus in which the male and female reproductive organs are segregated within the strobila, segments containing the male reproductive system alternating with those in which only the female reproductive system develops.

Examples: Taeniidae—*Taenia* L. (Fig. 29), *Echinococcus* Rudolphi

(Fig. 33); Davaineidae—*Cotugnia* Diamare, *Ophryocotyle* Friis; Anoplocephalidae—*Paronia* Diamare, *Hemiparonia* Baer; Dilepididae—*Paricterotaenia* Fuhrmann (Fig. 32), *Dipylidium* Leuckart, *Parvitaenia* Burt (Fig. 12L), *Onderstepoortia* Ortlepp (Fig. 12M); Hymenolepididae—*Hymenolepis* Weinland; Acoleidae—*Dioecocestus* Fuhrmann, *Diplophallus* Fuhrmann, *Gynandrotaenia* Fuhrmann, *Infula* Burt (Fig. 34), *Leptotaenia* Cohn; Amabiliidae—*Amabilia* Diamare.

3 *The Host as an Environment*

Guest and companion of the body,
into what places wilt thou now go.
HADRIAN

There are various associations of animals, but not all of them can be considered as parasitism. Other kinds of mutual relationships that may exist between the partners are known as phoresis, commensalism and symbiosis, and an understanding of these leads to a better grasp of the essentials in a parasitic mode of life. It may be that some cases of parasitism have arisen from phoretic, commensal or symbiotic associations, but it is generally considered that these must be few. However, an animal is a creature of opportunity and if there is a vacant environmental niche it does not remain untenanted for long, so that some parasites must have begun with a more benign relationship.

The associations discussed in this chapter which show these phenomena are not restricted to Platyhelminthes alone. The intention in so doing is twofold; to prescribe, as far as is possible, the different kinds of association that exist and to emphasise the universality of the phenomena elsewhere in the animal kingdom. The particular relationship between host and parasite is dealt with more fully in Chapter 11 on p. 120.

PHORESIS

An association between two animals where one of them is simply carried by the other is known as phoresis. One partner, indifferent to the association, is a mechanical carrier, while the other may be physiologically adapted to being carried about and cannot exist on a stationary *substratum*. Such a relationship is seen in the triclad turbellarian *Bdelloura* Leidy (Fig. 3) which lives on the gills of the king-crab *Limulus polyphemus* on the shores of New England, in *Ectoplana* Kaburaki on the Japanese *Limulus*, and in temnocephalids found in the gill-chambers of fresh-water crustaceans. These show adaptations which enable them to cling to their partners. They do not normally leave them, yet under certain conditions they can survive when removed, and in fact it has been possible to rear a temnocephalid (*Temnocephala brevicornis* Mont.) and maintain sexually mature and reproducing forms apart from the normal partners, provided there is an adequate supply of food and well-aerated running water. Phoresis is widespread in other groups for one has only to examine the pleopods of a water-louse (*Asellus*) to see numerous examples of associates that belong to this category. The pleopods act as respiratory structures and are in constant motion, ensuring adequate aeration. Accordingly, any organisms, such as protozoans (vorticellids and suctorians) and rotifers attached to them, are plentifully supplied with oxygen. The water-louse is apparently unaffected by these passengers and gains nothing from the association, but as the latter are unable to survive as free-living forms in still water they benefit from this association.

It has been argued that parasitism can originate in phoresis, but the temnocephalids which live on species of fresh-water crayfishes of the family Parastacidae do not support this view. The Parastacidae, or southern crayfishes, have a wide distribution in Australia, New Zealand, Madagascar and parts of South America and they all harbour temnocephalids on their gills. As the Crustacea which were ancestral to these modern parastacids date from Jurassic times and as the sea is an effective barrier to fresh-water Crustacea and to fresh-water temnocephalids alike, it would appear that parastacids and their phoretic temnocephalids were part of the fauna of the rivers of the great continental mass, Laurasia, before it split into the continents that we know today. The association of crustacean and temnocephalid may have started earlier but it could not have been later than the end of the Mesozoic epoch, and since then in the intervening millions of years this association has persisted as a phoretic one and has not developed into parasitism. There is however one known exception to this generalisation and that is the discovery, in Jugoslavia, of a temnocephalid *Scutariella didactyla* (Annand.) which lives as an ectoparasite feeding on the blood of a prawn.

COMMENSALISM

As the name suggests *commensalism* is an association of animals that feed together, or are messmates, though in fact the advantage, from a nutritional standpoint, is with but one of the associates. One of them may derive shelter, in addition to food, from the association, but if there is reciprocity and both partners benefit it is more usual to define the condition as *mutualism*. As it is frequently difficult to draw the line between commensalism and mutualism it is better to regard mutualism as a special case or as an extension of commensalism. A few examples from groups other than Platyhelminthes where this kind of relationship is more obvious will explain the phenomenon.

Hermit-crabs have soft bodies and nearly all of them use as portable shelters the otherwise empty shells of gasteropod molluscs. The commonest hermit-crab of the rock-pools of our shores, *Eupagurus bernhardus* (L.), can be seen with its abdomen tucked into the shell of a whelk or periwinkle and when alarmed the whole animal withdraws into the shell. Inside the same shell one frequently finds a bristle-worm, *Nereilepa fucata* Savigny, the two living in harmony. When the hermit-crab feeds, the bristle-worm leaves the upper whorls of the shell and, pushing its head between the mouthparts of the crab, snatches some of the food. This is a simple case of commensalism, the worm gets shelter and food but, so far as we know, gives nothing in return.

Hermit-crabs afford many other examples of commensalism and some of the associations are more complex. On most buckie-shells (*Buccinum*) inhabited by hermit-crabs there is a felt-like colony of hydroids, *Hydractinia echinata* (L.), but this association may be simply a phoretic one. On the other hand, if the coelenterate on the shell is an anemone such as *Calliactis parasitica* R. Q. Couch, which is frequently found below low-water mark, the association is definitely commensal, for the anemone benefits from particles of food which float up to it while the crab is feeding. There can be an advantage to the crab in this association for the stinging cells of the anemone give it some degree of protection while the anemone itself provides camouflage. One sees a greater advantage to the hermit-crab in another association of a different hermit-crab and a different anemone. The hermit-crab *Eupagurus prideauxii* (Leach) is associated with the sea-anemone *Adamsia palliata* (Boh.), these two always being associated and one never meets the one without the other. This hermit-crab lives in the rather globular shell of the mollusc *Natica* Scopoli, and when the young anemone first attaches itself to the shell it always does so in the same position, close to the mouth of the shell on the columellar side and as it grows it encircles the opening. The mouth of the anemone is directly under that of the crab and it feeds on the particles that fall from the crab's mouth: it is a living bib. This condition is commensalism, but the association does not remain static, for the anemone continues to grow and its foot eventually extends right round the last whorl completely encircling it and projecting in front of the shell-mouth. Anemone and crab grow together. The crab does not change its shell but continues to occupy it as well as the chamber formed by the foot of the anemone extending beyond the shell as a geometrical prolongation of the spiral, and a secretion of the anemone's foot forms an elastic lining to this living chamber. The reciprocal advantages here are that the hermit-crab gets continuous shelter and protection and the anemone has a continuous supply of food, so that from the initial commensalism there has developed a mutualism in which the partners are not only inseparable but are modified both in structure and behaviour.

Some polyclad Turbellaria are also commensals living with hermit-crabs. Two of them, *Emprosthopharynx opisthoporus* Bock and *Euprosthiostomum adhaerens* Bock, are associated with the hermit-crab *Petrochirus californiensis* Boucrez in the Pacific just as *Stylochus zebra* Verrill is found with *Eupagurus pollicaris* Say in the Atlantic. Many marine Turbellaria shun the light and are found during daylight under stones, within the intricacies of the roots of algae such as Laminaria, or, as in the case of some tropical forms, under rocks of dead coral. When such a rock is turned over the turbellarians, often present in considerable numbers, always glide away into such shelter as is available, or if removed into a glass vessel will invariably congregate in whatever part of the vessel is in the shade. The evolution of this negative phototropism may have contributed to the survival of these polyclads, and some of them, from finding shelter under stones, have found cover, shelter and a food supply in a shell occupied by a hermit-crab. Accordingly, it is reasonable to imagine that commensalism, at least in some polyclads, had its origin in the evolution of this character, and some would regard this as a first step towards parasitism.

Many bivalve molluscs (Lamellibranchiata) harbour numerous commensals and the Iceland-cyprina, *Arctica islandica* (L.), is no exception, for within its mantle-cavity there live commensally the pea-crab *Pinnotheres* Latr. and the rhabdocoel turbellarian *Urastoma cyprinae* (v. Graff). The latter is a small worm, at most barely one millimetre long, which, swimming by means of its cilia over the surface of the gills, feeds on the micro-organisms drawn by the clam into its mantle-cavity. When first discovered in *Cyprina*, as this clam was then called, *Urastoma* was thought to be the young stage of the large nemertean *Malacobdella grossa* O.F.M. which is also harboured by this clam but is thought to be a parasite. *U. cyprinae* and *M. grossa* are not restricted to one host, both being found in other lamellibranchs, but *U. cyprinae* is also found free-living in the boreal waters of the Atlantic. Under conditions which at present are not completely understood, *U. cyprinae* can become so numerous in the American oyster *Crassostrea virginica* (Gmelin) as to affect the oyster and apparently interfere with its food-supply. This then is an instance of a free-living organism that can live commensally and is in danger of being classed as an ectoparasite.

On the other hand some turbellarians play the major part in associations with other animals and act as hosts to commensals as well as to parasites. Some ciliates belonging to genera otherwise free-living, such as *Holophrya* Ehrb., are found within the gut of some planarians, while some species of *Trichodina* Ehrb. have the same external association with fresh-water planarians as others have with the polyp Hydra. A planarian can be the host of a stage in the life of a fluke, *Triganodistomum* Simer, which is parasitic in a fish, but this is discussed later.

There are many protozoans that are commensals and some of them find shelter and food within man's body. Man may find them living in his mouth or in his intestine, feeding on bacteria and organic debris, and the question arises again, can these commensals become parasites? One cannot generalise, but under certain conditions some commensals can adopt a parasitic mode of life. In the life-cycle of *Entamoeba histolytica* Schaudinn there occur small, non-invasive or commensal forms, and these can develop into the larger invasive and pathogenic forms, although the conditions bringing about the change are not known. Similarly, the ciliate *Balantidium coli* (Malmsten) is found as a harmless commensal in the gut of pigs, but if a man or a monkey became infested the ciliate would do extensive damage to the wall of the large intestine, causing a dysenteric balantidiasis. The change from commensal to parasite might be due to a lessening of topical restraint on the part of the host, or to the presence of some factor that induces the change. One can answer the question that we posed by stating that some commensals can become parasites and that this is evidently one of the ways in which parasitism had its origin.

ECTOPARASITISM

The state of a parasite living at the expense of its host, but remaining outside it, combining some of the phenomena seen in phoresis with some seen in endoparasitism is known as ectoparasitism. The parasite has adaptations which enable it to remain attached, preventing it from being dislodged and adaptations which enable it to live and feed on its host. The environment of an ectoparasite is a dual one. There is first the environment provided by the host itself. This is seen in the nature of the body to which the parasite is attached, such as the character of the skin or scales in cold-blooded animals, the warmth of the body of a bird or mammal, the presence of mucous glands, the thickness and nature of the subcutaneous layers and, of paramount importance to some parasites, the changing humoral composition of the host's blood or lymph on which the parasite feeds. These factors, provided by the host, are sometimes referred to collectively as the micro-environment, in contrast to the other factors, those not stemming from the host, referred to as the macro-environment. This can be the environment with which the host has to contend and it may be a changing one as regards temperature, humidity or salinity for example, or it may be the environment of the parasite when separated from its host, for there is a period in the lives of many ectoparasites when they are separated from their hosts.

The adaptations of ectoparasitic Platyhelminthes with regard to micro- and macro-environments, and in dissemination and transference from host to host, are dealt with later under Turbellaria and Monogenea where it will be seen that the general principles discussed here with regard to other ecto-parasitic forms also apply.

Ectoparasitism may be regarded as an extension of phoresis and that where opportunity arose advantage was taken of it. It is frequently difficult to decide into which man-made category we should place a particular associate unless we can determine whether it derives its food directly from its partner and whether the particular association is vital to it. Even so, we may get into difficulties as to whether a particular animal is an ectoparasite or a predator, difficulties concerning its physiology or our semantics. As a general guide to the phenomenon regarded as ectoparasitism a criterion could be whether a particular animal lives, feeds and is dependent on its partner. A few examples may help to clarify the position.

On most birds there are to be found biting lice which live through all the stages of their lives on the bodies of those warm-blooded animals while a few members of this group similarly infest mammals. The association is a close one for the lice have no independent existence. It is usual to regard these insects as ectoparasites for their presence in large numbers can cause considerable irritation although the food of most of them is restricted to dead tissues such as feathers or hair. On the other hand there can be no doubt that the sucking lice, an allied group that are blood-sucking and restricted to mammalian hosts, are true ectoparasites. Another group, the members of which are also ectoparasitic and blood-sucking, are the specialised flies known as Pupipara. Some lack wings, others with wings are poor fliers, but all are greatly modified and adapted to an ectoparasitic life. Included with them are the sheep 'ked' often referred to, although incorrectly, as a sheep 'tick', *Melophagus ovinus* (L.), nycteribiids which are ectoparasites of bats, *Braula* Nitzsch, an ectoparasite of bees, and the streblid *Ascodipteron* Adensamer, also of bats. Some of the winged forms shed their wings when they find a host, but *Ascodipteron* goes a stage further by burrowing under the skin and casting its legs as well as its wings, reaching an intermediate stage between ectoparasitism and endoparasitism, for its posterior end with the apertures of the respiratory system remains in communication with the exterior.

Fleas were formerly considered as predators and not as ectoparasites because, like mosquitoes, they did not spend the whole of their lives on their hosts and it was thought that they were only dependent on their hosts for an occasional meal of blood. However, as in the case of the streblids, some fleas such as the 'chigger' *Tunga penetrans* (L.) burrow through the skin of their host and become endoparasitic. The recent researches of Miriam Rothschild have demonstrated that fleas are as specialised and as intimately dependent, each on its particular host, as are endoparasites on theirs. This dependence is apparent in the effects produced in the rabbit-flea, *Spilopsyllus cuniculi* Dale, by changes in the factors of the micro-environment of its host, the rabbit. These factors, such as the raised temperature of the mating rabbits, the release of hormones in the doe, first by the pituitary following ovulation, and later in pregnancy by the adrenal cortex which initiates the production of corticosteroids in the blood, all affect fleas in their behaviour, metabolism and reproduction. A similar dependence of ectoparasites on their hosts has been shown in some monogeneans, which will be dealt with later.

Ectoparasitism of blood-sucking bugs does not go quite so far as it does in fleas. In this order of insects (Hemiptera), members of which possess piercing and suctorial mouth-parts which are used on plants and animals, it is not astonishing that some have departed from a life of predation to become ectoparasites. Assassin bugs (reduviids), although usually preying on other insects, occasionally attack man. One of them, *Reduvius personatus* (L.) frequents houses and preys on insects, including bed-bugs, and occasionally attacks man. This cannot be called ectoparasitism, but that condition is more nearly approached by the voracious blood-sucker *Triatoma* Laporte of South America which shows little discrimination in its selection of hosts, and consequently becomes a carrier of certain diseases common to man and other animals. Bed-bugs show greater adaptations to parasitism than either of these and there are the two kinds that affect man, *Cimex lectularius* L. in Europe and N. America, and *C. hemiptera* Fabr. in Asia. Other cimicids are ectoparasites of birds, including domestic poultry, and of bats. The most specialised bugs are polyctenids which are ectoparasites of bats. In fact, they are so highly modified that it is difficult to recognise them as bugs at all and few insects have caused more trouble to taxonomists. At first they were considered as aberrant diptera allied to the nycteribiids of bats, then they were classed with the sucking lice, and later transferred back to the Diptera as relatives of the cattle-flies (hippoboscids). Now they are thought to be bugs. But whatever their systematic position, they are veritable ectoparasites of false-vampires (*Megaderma* Geoffroy), large fruit-bats (*Cynopterus* Cuvier) and small insectivorous bats (*Taphozous* Oken) of the tropics and warmer regions of the Old World, and of mastiff-bats (*Molossus* Geoffroy) in tropical America. They bear the hall-marks of many ectoparasites of furry animals, which are flattened comb-like structures on the head, characters shared with many fleas, with sucking lice and with the specialised parasitic beetle *Platypsyllus castoris*, which lives on the Canadian beaver and looks more like a biting louse than a beetle.

ACCIDENTAL PARASITES AND PARASITES IN TRANSIT

Parasites found in hosts other than those normal to the species may be referred to as accidental parasites. There is a considerable number recorded from man, greater than the number found in any other animal, probably because man is more frequently subject to examination than are other animals, and because man may live in close association with so many animals, domestic and otherwise. As we have seen, a common commensal of pigs is the ciliate *Balantidium coli* (Malmsten), which, however, is frequently found as a parasite in swine-herds and others who work with pigs. Similarly parasites of dogs, cats, rats, sheep, cattle and even ducks have been recorded from man, and it is not surprising that some of the parasites of dogs, such as the dog-tapeworm, *Dipylidium caninum* (L.), are more common in children than in adults. A parasite in an alien host may be unable to survive in the different environment and thus fail to become established, or it may behave in an abnormal manner and wander about inside the body, as in *larvae migrantes*, seeking, to put it anthropomorphically, a suitable site or environment. But some parasites have pre-adaptations which enable them to become established in the different environment of a foreign host. As far as the parasite is concerned this may turn out to be a blind alley leading to extinction, either through the lack of those ecological factors enabling them, or their offspring, to be transmitted, or through killing the new host.

Parasites in transit are those that never become established, either because the environment is unsuitable or because they are present in a stage that is non-viable. A few examples will make this clear. A cosmopolitan parasite of many mammals, including rats and man, is *Capillaria hepatica* (Bancroft), a thin, fragile nematode that lives in the tissues of the liver and whose eggs accumulate there and do not pass out by way of the bile-duct and intestine. In the liver, development of the eggs proceeds as far as the 8-blastomere stage, but no further, and it is only when free of the body of the host, in an environment with an adequate supply of oxygen, that development can continue and give rise to a viable larva. Accordingly, the larva can only develop when the liver of the infested host is eaten by a predator and the undigested eggs are disseminated in its faeces, or when the host dies and its tissues disintegrate and are dispersed. Recovery of eggs from the faeces of man does not indicate infestation with *C. hepatica* but only that the eggs have passed through in transit. This is frequently the case in Central America where peccaries (*Dicotyles tajacu* (L.)), and their livers which are commonly parasitised, are eaten, and the eggs pass through man in transit and can subsequently complete their development. Another example is the part played by sea-gulls in the dissemination of the ox-tapeworm, *Taenia saginata* (Goeze). Sea-gulls may pick up on the shore segments of tapeworms that have arrived there from man in sewage and the cestode eggs may pass through their bodies unaffected as parasites in transit to be scattered on the land in the bird-droppings. When these are picked up by cattle they become infestive cysticerci. A final example is of a different order. The identification in human stools of eggs of nematodes which can infest man does not necessarily imply, as we have seen, that the worms producing the eggs are actually present in the body. In the same way plant-parasitic nematodes may be parasites in transit. A nematode, *Heterodera carotae* Jones, frequently present in carrots, does not become established in man, but if that vegetable is eaten raw the live eggs may be considered as parasites in transit.

FACULTATIVE PARASITES

Some organisms, plants as well as animals, can live both as free-living and as parasitic individuals, and these are called facultative parasites. It is a study of the environment of such forms and of their physiology that indicates one of the ways in which parasitism undoubtedly originated. Most facultative animal-parasites are either saprozoic or live in an environment rich in decomposing organic substances, i.e. rich in bacteria, on which they feed.

One example can be taken from the nematode family Rhabditidae in which are grouped free-living species as well as others parasitic in earthworms, molluscs, arthropods and even vertebrates. The genus *Rhabditis* Dujardin contains some species which have obligate phoretic stages in their life-cycles, others that are facultative parasites and yet others that are obligate parasites. It is thus on the border-line of parasitism. Encysted in the coelom or seminal vesicles, or free within the nephridia of earthworms are to be found larvae of *Rhabditis pellio* (Schneider). When the earthworm dies these emerge and feed on the decomposing body and so reach maturity. The adults reproduce and, so long as the medium is sufficiently rich in organic material, they may continue generation after generation as free-living individuals. If, however, food is inadequate, the larvae disperse to find another source, or to enter an earthworm, usually by way of its dorsal pores. Here they remain active, or encyst. *Rhabditis* has thus both a free-living existence and an endoparasitic one, although the latter is only part of the life-cycle. It is of interest to note that some birds can act as disseminators of *Rhabditis*, for the encysted larvae in the earthworms that they eat can pass through a bird's gut unchanged. This is yet another example of a parasite in transit.

The larvae of certain flies are also facultative parasites, since, like *Rhabditis pellio*, they can develop in organic matter which is undergoing bacterial decomposition. Many flies are attracted to, and lay their eggs in purulent wounds which results in an invasion of the tissues of the host by maggots, a condition known as *myiasis*. Most of these are species of the genera *Calliphora*, *Lucilia* and *Chrysomya* of R.-Desvoidy and *Wohlfahrtia* Brau. et Berg., belonging to the families Calliphoridae and Sarcophagidae and include the well-known blow-flies that attack sheep. These flies are facultative parasites and are in transition between the free-living and the parasitic modes of life. They lead on to the allied family Oestridae which comprises obligate parasites.

SYMBIOSIS

A number of different associations has been described as *symbiosis*, associations as different in nature as that of a hermit-crab and the anemone on its shell, or that between different castes of termites, or between a termite and the fungus *Penicillium* that it cultivates for food. If, however, we trace the word to its origin, we find that it was coined by de Bary in 1879 to describe associations such as that seen in lichens between fungus and alga that had then become known from the work of Schwendener. The word is now extended to include associations between plant and animal, or between animal and animal, in addition to the original one of an association between plant and plant. But the criterion of symbiosis is a relationship between associated organisms that is intimate and physiological, mutually beneficial and constant and permanent in character.

Even in animals as small as Protozoa we can find symbionts, for within the bodies of radiolarians there are present unicellular plants called *Zoochlorellae*, and the mutual benefit between these two organisms lies in the fact that the waste products of the radiolarian, carbon dioxide and nitrogenous substances are used by the *Zoochlorellae*, which in turn give off oxygen to the advantage of the radiolarian. Similarly, within the bodies of many coelenterates such as those that form coral, there are present minute, symbiotic algae, called *Zooxanthellae*, which function in a similar manner, and which association is ultimately responsible for the building up of barrier reefs and atolls.

Examples nearer to hand are the symbiotic relationships which exist between ruminants—sheep, cattle, deer—and the ciliates in their rumen, and again between horses and the ciliates in their caecum. These ciliates, belonging to the Oligotrichida, are naked except for a few cilia restricted to one end of the body where, specialised as membranelles, they are arranged in one or two spirals. They show elaborate intra-cellular or acellular specialisation in the development of organelles associated with nutrition such as a mouth which lies in the centre of one of the spiral membranelles, gullet, stomach and even rectum. There is a neuromotorium that controls the membranelles and a double nuclear structure, common however to all ciliates, and even in some a structure described as a skeleton but apparently associated with the storage of carbohydrate. Some of these highly specialised forms, such as species of the genus *Diplodinium* Schuberg and *Polyplastron* Dogiel, are obligate anaerobes that can digest cellulose. In the rumen they feed on cellulose eaten by their 'host', thrive and multiply, dividing about once every twenty-four hours, so that there is a daily turnover to the 'host' of animal protein equal to the bulk of the average number of protozoa present. Estimates of the number present vary, but one for *Diplodinium* in the rumen of sheep is 240,000 to 450,000 individuals per ml, and *Diplodinium* is but one of a number of genera. There are, likewise, estimates of the amount of food that these ciliates make available to their 'hosts' but it is of interest that the estimates of well-nigh a century and a quarter ago, of one-fifth of the animal's requirements, made by Gruby and Delafond who discovered this relationship, do not differ significantly from present-day estimates. In this type of symbiotic relationship one of the partners, the ciliate, receives from the host all its food in a liquid, warm, and methane-saturated environment in which it can live. The other partner, the ruminant, unable to digest cellulose itself, has a part of its cellulose digested for it by ciliates and actually receives the products of fermentation of carbohydrates—acetic acid, butyric acid, lactic acid—in addition to the protein from the ciliates it digests. An analysis of the carbohydrases of one of the larger ciliates of the rumen of sheep, namely *Polyplastron multivesiculatum* (Dogiel et Fedorowa) (Fig. 13B) has been made possible by experiments which established this ciliate as the only large one of its kind present in the rumen. The ciliate could thus be obtained in sufficient quantity for a full and detailed biochemical analysis. This showed that not only are cellulose and cellobiose hydrolysed, but many other carbohydrates including amylose, maltose, xylobiose and pentosan are also broken down. In this instance the mixed bacteria, which were also present in the rumen and prepared in the same way as the ciliates, did not affect the breakdown of cellulose.

A symbiotic phenomenon, somewhat analogous in nature, is seen in the association between many termites such as *Kalotermes* Hagen and flagellates such as *Trichonympha* (Fig. 13A) which live in a hypertrophied section of

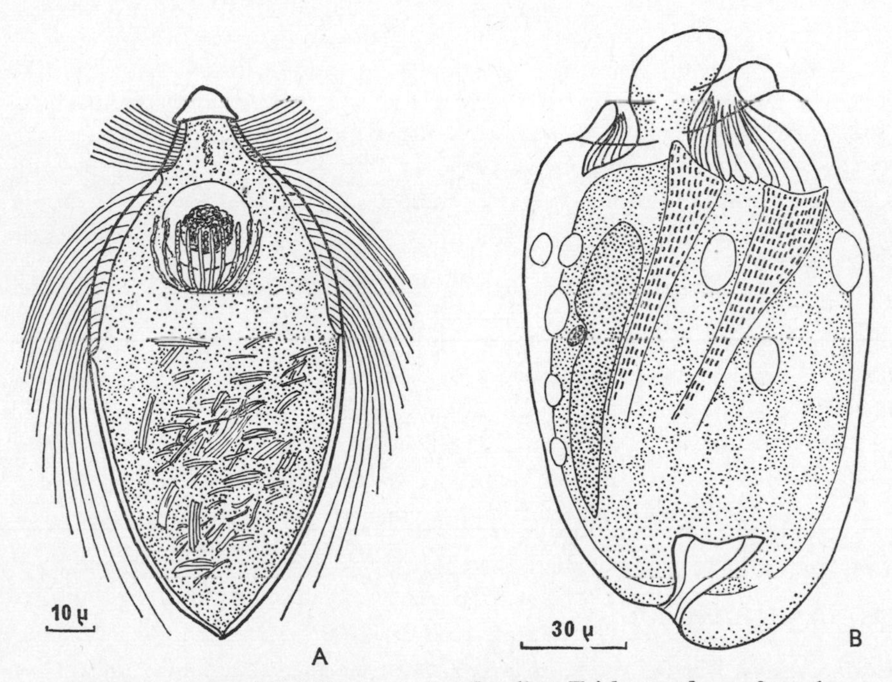

FIGURE 13 Symbiotic Protozoa. A, a flagellate *Trichonympha* sp. from the gut of a termite showing ingested particles of wood; B, a ciliate *Polyplastron multivesiculatum* from the rumen of sheep. (After Kofoid.)

their hind gut. The termite, in virtue of its protease, is able to digest the protein contained in wood, but it has no cellulase with which it can digest the bulk of its diet. This, however, is done by the symbiotic flagellates which actually ingest the particles of wood in the termite's gut. It would seem that the relationship between flagellate and termite is that the former, provided with shelter, a suitable anaerobic environment and food by the latter, hydrolyses cellulose to glucose and subjects the glucose to anaerobic fermentation. The products of this digestion are then utilised by both flagellate and termite.

PARATENIC HOSTS

A paratenic host is a very special case of an intermediate host in the transit of a parasite, that is an optional and additional one to those hosts essential to the life-cycle. The phenomenon was first recognised by Joyeux and Baer who used the term *hôtes d'attente* which could be translated as 'waiting hosts' or 'temporary hosts', but the English translations were not the precise equivalent of the French. To remedy this Baer replaced the term with *paratenic*, a word that introduces another aspect of the concept, that of time in the life-cycle in prolonging or extending it. A paratenic host is always an intermediate host of a larval stage, but a stage in which development does not occur. An example can be seen in the life-cycles of pseudophyllidean tapeworms such as *Diphyllobothrium* Cobbold (Fig. 30) parasitic in fish-eating birds or mammals in which life-cycles there are three larval forms, *coracidium*, *procercoid* and *plerocercoid*. The third larval stage, the plerocercoid, develops in a fish and encysts in part of its body. The mature, segmented tapeworm can only be formed when the fish is eaten and the plerocercoid can reach the gut of a warm-blooded or homoiothermic animal. But before reaching the final or definitive host the plerocercoid may pass through one or more paratenic hosts in the form of other and predator fishes. A small fish harbouring a plerocercoid may be eaten by a larger fish in which it becomes re-encapsulated, and this fish may in turn be eaten by a third and so on. Each of these subsequent fishes receiving and maintaining the plerocercoid is a paratenic host.

ENDOPARASITISM

Endoparasites live inside their hosts and are, without exception, completely dependent on them, while their hosts are in no way dependent on their parasites. These characters separate endoparasites from all other associations. In all parasites there is a principal or definitive host in which maturity is attained and in which reproduction takes place, and there may, or may not be, throughout the parasite's life, free-living stages and stages in intermediate hosts. The parasite may thus have to contend with a great number of different environments and one could select a whole series of examples to illustrate the number and variety of these.

First there is the simplest case, and actually among the rarest, where the parasite is never in an intermediate host and never free-living. This is seen in one of the trypanosomes of horses, *Trypanosoma equiperdum* Doflein, which has one environment, that of the horse, and which passes by direct contact during coition from stallion to mare or *vice versa*. Secondly, there are parasites, such as the pin-worm *Enterobius vermicularis* (L.), living in the gut of man whose only other environment is that of the egg when outside the body, in which stage it is transmitted to another individual. Thirdly, there are parasites with a free-living stage, such as the hook-worm of man, *Necator americanus* (Stiles), the adult of which lives in the gut, while the larva has first to contend with the external environment before entering the human body, and then within the body changes in environment as it moves from one region to another before reaching the intestine. Fourthly, where there is an intermediate host, as in the case of the ox-tapeworm, *Taenia saginata* (Goeze), there are two main environments, the gut of man and the muscles of cattle, ignoring the environment of the egg outside the body and the different environments encountered inside the ox before the parasite reaches the muscles. Lastly, we can take an example of a parasite with both a vertebrate and an invertebrate host, involving a large number of different environments, although by no means the greatest number that can be encountered. The great complexity in the life-cycle of some parasites is seen in that of the fluke, *Parorchis acanthus* Nicoll (Fig. 19). This fluke lives in the rectum, or in a diverticulum of the rectum called the *bursa Fabricii*, of the herring gull, *Larus cinereus* Brisson. Its larva, a miracidium, which is born there, passes through three environments, the hind gut of the gull, the sea in which it is free-swimming and that of the body of the whelk, *Nucella lapillus* (L.) which it penetrates. Within the environment of the digestive gland of the whelk there are various developmental stages, but the larva that eventually emerges, the cercaria, has again the sea as its environment and then various alternative hosts in transit, on or in which it encysts, as on the gills of the mussel, *Mytilus edulis* L. When the mussel is eaten by a gull the young fluke traverses the whole length of the gut with its changing environments before reaching the *bursa Fabricii* where it reaches maturity.

The various factors that determine the environment of an endoparasite are temperature, oxygen availability or tension, acidity or alkalinity (pH), osmotic pressure, oxidation-reduction potential (Eh), the specific inorganic and organic substances present and the mechanical effect of the host's tissues on the parasite. These vary from site to site and even within the same organ. The commonest site for parasites is the gut, for the obvious reason that the majority of parasites enter the body by the mouth, intermediate stages of many parasites being on or in the food eaten. Other regions are: the lungs and other parts of respiratory organs except gills which are usually regarded as an external site; the heart, blood-vessels and the haemocoel; the various glands associated with the gut, such as salivary glands, liver and pancreas and their ducts; the kidneys, urinary bladder and their ducts; the coelom; the nervous system including the brain and the eye itself; lymph glands and lymphatics; reproductive system and ducts; connective tissue and fat; muscles; and even the inside of bones. There is not a single organ or tissue in one animal or another that is not the site of some parasite, and no two sites possess the same environment, while the environment may even vary from one part of an organ to another. This is obvious in the case of the alimentary canal in which, in some animals, the anterior region of the gut may be slightly alkaline, the stomach strongly acid while the acidity is reduced in the mid-gut and hind-gut to near neutrality. The inorganic and organic substances derived from the breakdown of the food of the host, which are available as food for the parasite, also vary along the length of the gut. There may be no oxygen available in the centre of the lumen of the gut, but a small quantity available in the mucus in contact with the gut-wall or even more if the parasite is attached and feeds on blood obtaining some oxygen from the haemoglobin in it. The different environments in different regions of one site are less obvious in an organ such as the liver, although here there is a difference in the environment of the hepatic cells from that of the bile in the bile-ducts within the liver or from that of the bile in the gall-bladder.

Temperature too plays its part in the climate or environment of the parasite. Temperature may change in relation to external temperature as it does in poikilothermic and heterothermic animals, but is much more constant in homoiothermic animals. The dependence of a parasite on a certain degree of warmth is seen in the malarial parasite, *Plasmodium* March. et Celli, which cannot survive at low temperatures. This accounts for its absence from regions where the temperature falls below 20° C, a temperature not inimical to its host the anophelene mosquito. Temperature is one of the changing factors in the environment of a parasite when it passes from an invertebrate intermediate host to a homoiothermic one. A plerocercoid, as for instance that of *Ligula intestinalis* (L.), which has remained encapsulated for months in a fish, is taken into the gut of a bird or mammal. There, under the influence of the altered environment, including a higher temperature, it becomes a strobilate worm, reproduces and completes its life-span in a matter of two or three days. This is understandable, for the sudden rise of the temperature from that of a fish to that of a bird must have a marked effect on the metabolism of the parasite. Another instance where there is a rise in the ambient temperature is seen in its effect on the development of, and infestation by, some parasitic copepods of cyprinid fishes, shown in the case of *Lernaea cyprinacea*. Temperature also controls the emergence of cercariae of *Renicola* from the periwinkle *Turitella*.

Even in a warm-blooded mammal the temperature varies from one part of the body to another, and we cannot imagine that it is other than temperature that causes the guinea-worm, *Dracunculus* Reichard to behave as it does. Living in the subcutaneous tissue of man it comes to the surface of the body and liberates its aquatic larvae from an ulcer which it induces, as in the leg of an Egyptian agricultural worker while working in water, or on the back of an Indian water carrier, each site being the cooler on account of contact with water.

Once the parasite is established in a particular site in an organ, one would imagine that the environment would remain more or less static, but this is by no means the case, for the environment can alter with the nature of the food of the host, while the mechanical presence of the parasite or the production by it of toxins can cause a reaction in the tissues of the host again with consequent alteration in the environment. A result of this can be seen in an effect on the parasite, which limits its powers of reproduction, and results in either *numerical restraint* or *topical restraint*, the restriction of the parasite to a limited region of the body of the host. Moreover the altered environment may be inimical to other parasites of the same or related kinds and be effective in preventing further infestation, a condition known as *premunition* discussed more fully under host-parasite relations. Premunition persists only so long as there are parasites present, the resistance to further infestation ceasing when the parasites are expelled or otherwise lost. Premunition is seen in *Taenia solium* L., the pig-tapeworm or solitary tapeworm. It is well named *solium* for it is rare to find more than one present, and if there are more than one they are all of the same size, having resulted from a multiple infestation. A more lasting alteration in the environment of the same kind results in *immunity*, a condition which persists after the parasites have left the body of their host. This type of immunity is acquired immunity, a term used to describe the change induced by the environment in contradistinction to natural immunity, a quality inherent in some animals. Some parasites never induce an immunity but others do, and among the latter is a common nematode parasite of sheep, the trichostrongyle *Haemonchus contortus* (Rudolphi) which is found in the abomasum, or fourth stomach, of sheep and other ruminants. Like the hook-worm of man, *Necator americanus* (Stiles), this trichostrongyle has free-living larval stages. The third-stage larva is picked up by sheep in grazing and it reaches maturity in the abomasum where it becomes attached to the gut-wall and feeds on blood. The toxins which it produces have a profound effect on the host and a heavy infestation may cause death, but in a less severe infestation the response induced in the host results in an altered internal environment unfavourable to the parasite, which persists even in the absence of the parasite. Further infestation is thus not possible. One might note a practical application of this for it has been found that infestive larvae that have been irradiated with X-rays or otherwise rendered sterile evoke in sheep the same immunity reaction as that produced by normal infestive larvae. Accordingly, lambs that are given irradiated larvae in an amount not large enough to be lethal or cause a severe pathological effect acquire an immunity without becoming disseminators of the parasite. After being treated in this way they can resist a natural infestation when exposed to it in the field.

While relying entirely on their hosts for all their requirements, some parasites still retain the enzymes necessary to digest their food; others, however, are dependent on their hosts for certain essential food and growth-substances. As is seen in some protozoa there are degrees of dependence, and the greater the degree of dependence the more intimate is the relationship between parasite and host. There are endoparasites that are influenced by their host's hormones and reproduce in harmony with their host. A classical illustration of this is seen in *Polystoma integerrimum* (Fröhlich) (Fig. 14) from the bladder of the frog, *Rana temporaria* L., which does not reproduce until its host does at the age of three years. As a result the infestive larva of *Polystoma* is free-swimming at the same time as the tadpole with internal gills, the optimum stage for infestation, and the two life-cycles are thus synchronised. This is a dependence similar to that seen in some ectoparasites, as has been already noted in the case of the rabbit-flea.

In most cases of endoparasitism host and parasite live together in a steady state, the parasite taking its toll of its host's food and body and the host responding in the repair of its tissues and in the exercising of numerical and topical restraint. This state may, however, become unsteady and altered in one direction or another, the extreme result being either the death of the parasite or of the host, both of these being generally disadvantageous to the continuation of the race of the parasite. Equally disastrous is parasitic castration, a phenomenon first shown by Giard to obtain in Decapoda infested with various crustacean parasites such as *Sacculina carcini* Thomp. Parasitic castration is apparently not restricted to Crustacea for J. S. Scott has discovered that the intermediate stage of the tapeworm *Paricterotaenia paradoxa* (Rudolphi) (Fig. 32), which is a *polycercus*, causes castration in the intermediate host, the earthworm *Allolobophora terrestris* (Savigny).

OBLIGATE PARASITES

Ectoparasites and endoparasites that are incapable of living other than parasitically are obligate parasites. The parasite may be obligatorily parasitic throughout the whole of its life in one or more hosts, as in the case of trypanosomes and malarial parasites, gapeworms and tapeworms, or it may be obligatorily parasitic only as an adult, and free-living as a larva, as seen in nematodes such as hook-worms and trichostrongyles. The opposite may obtain, the adult being free-swimming, as found in monstrillid copepods, an example of which is *Monstrilla helgolandica* Claus, the gasteropod mollusc *Odostomia scalaris* MacG. being host for the obligate, parasitic larvae. There are also parasites in which the parasitic stages are separated by free-living or free-swimming stages as is common in the life-cycles of most flukes and in parasitic copepods, such as *Lernaea asellina* L. In all these parasites the parasitic condition is, by necessity and without option, obligatory.

Recent advances in our knowledge of the metabolic requirements of parasites has made it possible to culture in the laboratory, *in vitro*, some parasites that are otherwise obligate. In view of this exception and at the risk of pedantry, the definition of obligate parasites should state that 'in nature' they are incapable of living other than parasitically.

4 *The Parasitic Turbellaria*

Everything that lives, lives not alone,
nor for itself.
WILLIAM BLAKE

Turbellarian stock has already been named as that which gave rise to the different classes of Platyhelminthes, and it is to be expected that within the group there will be some indication of this. The turbellarians inhabit many different niches, free-swimming forms in the sea, others lurking under stones or among seaweeds in rock-pools, some in fresh-water streams and ponds and others in damp, terrestrial habitats. There are commensals in the digestive tracts and body-cavities of molluscs, echinoderms and crustaceans, there are phoretic forms, there are ecto- and endoparasites, and there are some that are apparently equally at home in fresh water, brackish water or the sea.

The parasitic Turbellaria are less numerous than the free-living forms and so it would appear that the most successful parasites derived from them have already evolved into those groups that form the other classes of Platyhelminthes, while the less successful, or perhaps those later in turning to a parasitic existence, are not yet sufficiently modified to be classed as separate groups. A few examples show the main features of the parasitic forms. The marine triclad *Micropharynx* Jägerskiöld is an ectoparasite on skate, differing from free-living Maricola in the loss of eyes and reduction in the ciliated covering. *Hypotrichina* Calandruccio, an allied species, is an ectoparasite living under the carapace of the crustacean *Nebalia* Leach. Here it is attached by a sucker at its anterior end and by its mouth with its powerful sucking pharynx which opens at the posterior end. It also shows a reduction of cilia which are restricted to a median strand on its ventral surface.

Many phoretic and ectocommensal forms follow the pattern of parasitic forms in that they are characterised by the loss of pigment, reduction of the eyes and of the cilia accompanied by a development of adhesive organs. In the ectocommensal forms the same loss of cilia is not always apparent although the eyes are frequently reduced while an increase in the size of the ovaries may be related to the availability of food.

One can see throughout the Turbellaria various morphological characters which foreshadow those of the other classes. This is particularly obvious in the case of the adhesive organs. Gland-cells which produce a sticky secretion can be seen in free-living, commensal and parasitic Turbellaria and also in the parasitic forms in other classes, as in the prohaptor of some monogeneans, the funnel of *Gyrocotyle* (Fig. 11) and the gland-cells of some digenean *miracidia* and *cercariae*. Muscles associated with glands, as for example the glandulo-muscular organ of the cave planarian *Kenkia* Hyman, are a feature of many turbellarians as they are of monogeneans. Muscular suckers are found in both these groups as well as in other classes of the Platyhelminthes, while hooks augment the action of suckers in some turbellarians as they do in many cestodes and in the opisthaptor of some monogeneans. It is not suggested that organs of attachment, so characteristic of the parasitic classes of the phylum, have been derived from those of the Turbellaria, or even that they are homologous, but it is noteworthy that these organs (glands, suckers, hooks and so on) have their counterparts in many free-living Turbellaria. One can only assume that there has been independent evolution of organs of attachment in those classes which share their ancestry with that of the Turbellaria.

There are other aspects of the Turbellaria that have their counterparts in the other classes. In asexual reproduction, as in some rhabdocoels, planarians and terrestrial triclads, transverse fission results in a chain of zooids, a phenomenon similar to that seen in cestodes, where perhaps there is a more precise parallel in *Haplobothrium globuliforme* Cooper, a tapeworm of the North American fish, the bow-fin, *Amia calva* L. In this worm, in addition to the division or budding which gives rise to the chain of segments forming the strobila, a further division of these segments and of the strobila results in a number of strobilae. There is a difference here, for whereas in the Turbellaria all the worms arising from a chain of zooids are alike, in *Haplobothrium* the scolex of the original strobila differs from the pseudo-scolices of the strobilae arising from it. In sexual reproduction we may also draw a parallel, for in sexual reproduction of the hermaphrodite rhabdocoels, as in the triclads, mutual insemination is the rule, as is the case in some monogeneans, a notable example being the monogenean *Diplozoon* v. Nordmann (Fig. 15), where two hermaphrodite individuals become permanently united.

In the order Eulecithophora, formerly called rhabdocoels, the group

Dalyellioida includes more species that are commensal and parasitic than other groups of equal taxonomic value, and noteworthy amongst these are species of the aberrant genus *Fecampia* Giard, species of which are parasitic in crustaceans. One is found in the body-cavity of the common crabs and hermit-crabs of our shores, another lives in the marine isopod *Idotea* Fabricius of northern waters, while a third lives in the large, flat, Antarctic isopod *Serolis* Leach from the waters of the Patagonian continental shelf. As *Fecampia spiralis* Baylis becomes mature within *Serolis*, the development of the reproductive organs is accompanied by the development of cystogenous or nidamental glands below the epidermis and the almost complete degeneration of organs other than the reproductive system. The sexually mature *F. spiralis* leaves the body of its host and secretes round itself on the dorsal plates of its host a flattened, spirally shaped cocoon. A curious feature of this cocoon is that the direction of the spiral is related to the site on the body; if on the left side the spiral is sinistral but is dextral on the right side. Within the cocoon *Fecampia* lays its eggs and apparently dies.

5 *The Monogenea*

There is a friend that sticketh closer
than a brother.
Proverbs xviii. 24

The monogeneans, usually considered as the ectoparasitic group of Platyhelminthes, are found on aquatic vertebrates (bony fishes, elasmobranchs, chimaeras, amphibians and reptiles, and one mammal) and on a few invertebrates (cephalopods and parasitic Crustacea). The larva or *oncomiracidium*, on eclosion, swims to its host, attaches itself by its larval hooks and grows to maturity developing the adhesive organs of the adult which are very characteristic bodies. The adults are found occupying various sites; some remain on the external surface, lying flat against the body and thereby offering little resistance when the animal swims through the water, others are found in the mouth or branchial chamber attached to the lips, palate, tongue, gill-rakers, or to the gills themselves, held in position against the current of water by clamps, hooks or suckers. A few appear to seek a more sheltered niche as seen in *Calicotyle kroeyeri* Diesing which is found in the cloaca of the thornback (*Raja clavata* L.) or *Chimaera* L. or in *Acolpenteron ureteroecetes* Fisch. et al. which has found its way into the ureters of the small-mouth bass, *Micropterus dolomieui* Lac., or *Gymnocalycotyle inermis* (Woolcock) in the oviducts of the false saw-fish *Pristiophorus cirratus* Lath. *Dictyocotyle coeliaca* Nybelin has invaded the body-cavity of skates, and *Amphibdella torpedinis* Chatin, whose normal habitat is on the gills, has been found in the bloodstream of *Torpedo ocellata* Raf. and *T. marmorata* Risso in which some development involving protandrous juveniles takes place. *Polystoma integerrimum* (Fröhlich) (Fig. 14) is well known as an endoparasite of toads and frogs, living in the bladder, but it starts its parasitic life attached to the gills of a tadpole.

ADHESIVE ORGANS

Wherever monogeneans are found there is seen to be a very high degree of specialisation of the organs of attachment and so powerfully is the monogenean attached that in some instances any attempt to pull one off results in the animal being torn in two. Some become permanently attached, while others move over the surface of the body or change their position from one part of the gill to another. Such has been noted in the case of *Dactylogyrus solidus* Achmerov, a parasite on the gills of the common carp, *Cyprinus carpio* L., which, when the oxygen tension in the water is low, moves to the place of optimum oxygen supply, the ventral ends of the first and fourth gill arches. There are others, however, that maintain their position in similar circumstances.

Monogeneans are attached by adhesive organs which are developed at both ends of the body, and as they are frequently not in the form of acetabula or suckers they are referred to as haptors, prohaptor at the front end and opisthaptor at the rear.

The reactions of hosts to the presence of these parasites vary. In some there is little damage to the host's tissues and little reaction to their presence and this is more commonly the case where the haptor consists of clamps such as those of *Diclidophora* (Fig. 5L, M). But the presence of dactylogyrids, whose haptors are armed with hooks, may result in haemorrhage and pathological growth. In the latter case the host's tissues may grow around the haptor as can be seen on the gills of the gadoids *Molva dipterygia* Sm. and *M. byrkelange* Ström when infested with *Linguadactyla molvae* Brinkmann.

The haptor is the very life-line of a monogenean, for if contact with the host is broken and it comes adrift, the life of the parasite is endangered. In the great variety of haptors that has evolved it is apparent that the operation of natural selection has been rigorous and survival has depended on the efficient attachment of parasite to host. It is possible, by examining a whole range of haptors, to see something of the course of evolution and of the adaptations to different environments of the host. This approach is not without its dangers for evolution does not always operate in such a progressive way.

The opisthaptors show great variation, but the most generalised is likely to be that which differs least from the larval one, a disc with sixteen hooklets. Such a haptor is seen in *Gyrodactylus* v. Nordmann (Fig. 5B) where, in addition to sixteen hooks arranged around the periphery of the disc, there is present on the disc a pair of large anchor-hooks, a feature common to many haptors. In *Udonella* Johnston (Fig. 5C), an ectoparasite of ectoparasitic copepods, there is complete absence of hooks and the disc becomes a sucker. In *Acanthocotyle* (Fig. 5E) there is developed, in front of the haptor, a large muscular disc bearing concentric rows of hooks, an arrangement similar to that seen on the suckers of cestodes such as *Davainea* and *Raillietina*. Another line of evolution is the perfection of the disc as a sucker, such as is seen in the entobdellids (Fig. 5D), where the efficiency of the sucker is increased by the development of a thin flange, projecting beyond the margin, which can be applied closely to the irregularities of the surface of the host's body. From the specialisation of the disc as a single sucker, specialisation in another direction is seen where the surface is divided by a number of ridges radiating to a marginal ridge thus forming a series of loculi or smaller suckers as seen in *Trochopus* Diesing and *Monocotyle* Taschenberg, where too the mechanism is augmented with anchor-hooks. A further development in this direction is an increase in the number of loculi with a division of the discs to give a central loculus and a series of concentric loculi around it, an arrangement like that in the haptors of *Merizocotyle* Cerfontaine (Fig. 5G) and *Tristoma* Cuvier, or in the greater specialisation of a few loculi into more elaborate suckers. This latter condition is that of *Polystoma* (Fig. 5H) where the opisthaptor is a disc with six well-developed suckers, augmented by anchor-hooks. In the opisthaptor of *Oculotrema* Stunkard (Fig. 5I) there are six suckers of a more specialised kind, but anchor-hooks are wanting. Six suckers are also seen in hexabothriids (Fig. 5N) where each of the six suckers is provided with a hook within it, and the opisthaptor ends in an appendage armed at its tip with an additional pair of suckers.

From the sucker it is but a step to the clamp. The clamp is sucker-like but the development of sclerites in its walls and around its margin limits its movement to the apposition of one side of the lip to the other. In most species the number of clamps is four pairs, but the number is not invariable. As in the case of suckers some are sessile, like those of *Diplozoon* Nordmann, and some pedunculate as in *Cyclocotyla* Otto. In the pedunculate forms the presence of spines on the outside of the clamps, as in *Diclidophora* Diesing, can effect a grip between adjacent clamps. Another development is an increase in the number of clamps which can even extend along the margins of the body, or in the great development of the anterior pair apparently at the expense of the remainder which are small and pedunculate; the former condition is a character of *Microcotyle* v. Beneden et Hesse (Fig. 5P) and the latter of *Anthocotyle* of the same authors.

In the opisthaptors so far considered development has been that of the disc into loculi, suckers and clamps, with anchor-hooks augmenting the adhesive action, but evolution in another direction has resulted in the specialisation of the anchor-hooks to form a different type of organ of attachment. Marginal hooks are present, sometimes fewer than the sixteen characterising the larva, while in the centre of the disc are the anchor-hooks, two pairs of which are present in *Amphibdelloides* Price and *Tetraoncus* Diesing (Fig. 50). They are connected together by supporting bars and one pair projects dorsally and the other ventrally. A mechanism of muscles and tendons enables these hooks to be protruded or withdrawn and functions in attachment, or in detachment when the parasite changes its location on the gill-lamellae. In diplectanids, shield-like discs covered with scale-like spines are present on the dorsal and ventral surface of the disc in front of the anchor-hooks, and these bodies, termed squamodiscs, function in connection with the anchor-hooks.

In addition to these organs of attachment there are in association with the disc the less obvious but nevertheless efficient dermal glands of a glutinous or sticky nature, reminiscent one might add of the posterior glands of some rhabdocoele turbellarians.

The anterior end of the body also shows in the prohaptor certain organs of attachment, and although mainly concerned with attachment of the mouth during feeding they also function during locomotion. This concerns two regions: the mouth with its pharyngeal and buccal suckers and the anterior attaching bodies in the form of bothria, suckers, pits and papillae. They are less conspicuously noteworthy than those of the opisthaptor, yet as in the latter they show in their origin and development a natural selection and specialisation.

Another adaptation in the lives of monogeneans is seen in the correlation between lives of hosts and parasites, a relationship first clearly shown by Zeller in the life-cycle of the common frog and that of its parasite *Polystoma* Zeder, but now realised to be but one manifestation of a phenomenon of wider occurrence.

CORRELATION BETWEEN LIFE-CYCLES OF HOSTS AND PARASITES

Polystoma integerrimum (Fröhlich) (Fig. 14) lives as a parasite in the bladder of the common frog, and produces eggs at the time of the frog's spawning, and on eclosion the larval *Polystoma*, the *oncomiracidium*, becomes attached to the internal gills of a tadpole, where it lives as an ectoparasite. When metamorphosis of the tadpole takes place, which is accompanied by the resorption of the gills, the young *P. integerrimum* passes down the gut of the young frog to the urinary bladder where it lives endoparasitically, becoming sexually mature at the end of three years at precisely the same time as its host. If, by chance, the tadpole-host is a very young one still possessing external gills, as occasionally happens, the parasitic, larval *Polystoma* develops rapidly into a neotenic individual, and gives rise parthenogenetically to a few eggs. These soon hatch, the oncomiracidia become attached to the internal gills of more mature larvae, and develop into normal individuals. The frog takes to water

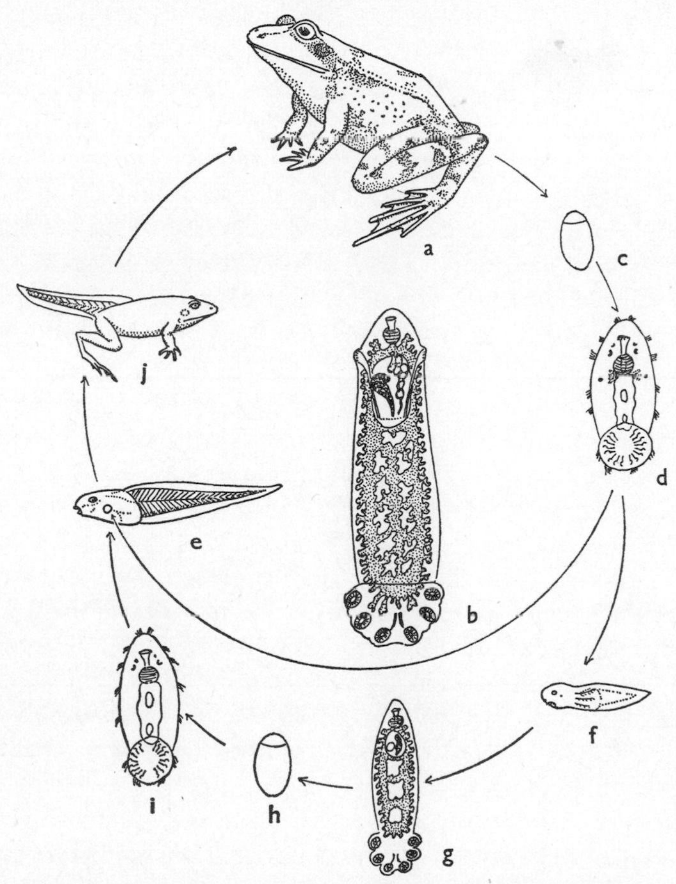

FIGURE 14 Life-cycle of *Polystoma integerrimum* (Fröhlich).
The definitive host, a frog, such as *Rana temporaria* L. (a) harbours *Polystoma* (b) in its urinary bladder. Eggs are released in spring when frogs enter the water to breed, and from each egg (c) there emerges an oncomiracidium (d) which enters the spiracle of a tadpole with internal gills (e) and attaches itself to the gills. Alternatively, if the oncomiracidium (d) encounters a tadpole with external gills (f) and attaches itself to these it develops rapidly into a neotenic individual (g) which produces a few eggs. From each egg (h) there hatches an oncomiracidium (i) which in turn seeks the internal gills of another tadpole. During metamorphosis of the tadpole (j) the juvenile *Polystoma* passes down the gut and enters the developing urinary bladder.

in the spawning season and it is then that *Polystoma* produces eggs, while it is only at the end of three years that both host and parasite become sexually mature. This life-cycle is not unique, for in the bladder of the common tree-frog *Hyla versicolor* (Leconte) and the squirrel tree-frog *H. squirella* Lat., both of North America, one may find an allied species *Polystoma nearcticum* Paul. The life of this *Polystoma* follows a similar pattern in that a neotenic individual develops when the oncomiracidium attaches itself to the external gills of a very young tadpole. But when the oncomiracidia produced by the neotenic form, or those hatched from eggs of parasites in the bladder, encounter a tadpole undergoing metamorphosis, they enter the cloaca and in this way reach the bladder, where they develop to maturity in harmony with the sexual development of their host. *P. ozakii* Price also seems to behave in the same way because larvae in which the first pair of suckers had not developed have been found in the Asiatic frog *Rana chensinensis* Dov. in Sakhalin. Synchronisation of life-cycles could not be closer and it has been shown that it is the gonadotrophic hormones of the frog that influence the sexual development of its parasite.

A similar phenomenon is seen in the life-cycle of the monogenean *Mazocraës alosae* Hermann and its relation to its host, various species of shad of the genus *Alosa*, e.g. *Alosa brashnikovi* (Borodin), *A. caspia* (Eichwald) and *A. saposhnikovi* (Grimm), found in the Caspian Sea. These fishes, living for the greater part of the year in the open sea, do not as a rule spawn until they are three years old. It is then that they approach the shore and ascend rivers to spawn during a short period at the end of May and the beginning of June. At this time *Mazocraës* lays its eggs which are found in large numbers attached to the gills of the shad. They hatch in about eight days and the infestive larvae become attached to other fishes. As the young shads do not hatch until after the mature shads have returned to open water they do not have contact with mature forms until they themselves are mature at the age of three years and the absence of parasites on one- and two-year-olds is therefore readily understood. As gravid parasites are never found unless on shads about to spawn there may here be an influence of the hormones of the host on the maturation of the germ-cells and oviposition of the parasite similar to that seen in *Polystoma*.

In the closely related parasites of the Japanese mackerel *Scomber japonicus* (Houttuyn) there is a similar phenomenon. These parasites *Kuhnia scombri* (Kuhn) and *Kuhnia minor* (Goto) are found only on two-year-olds and older fish, for infestation occurs when they move into shallow water to spawn, when the reproduction of the monogenean and the spawning of the mackerel are synchronised.

DIPLOZOON PARADOXUM

There are many curious features in the life-cycle of *D. paradoxum* v. Nordmann which has already been referred to as an instance of reciprocal fertilisation (Fig. 15). About a dozen and a half different forms of *Diplozoon* have been recorded to date in different parts of the world from different fishes, and in most cases the specificity appears to be the restriction of one species of parasite to one kind of host. *D. paradoxum* is the exception for it is recorded from over thirty different kinds of fish. However, when one considers that until 1950 only three species of *Diplozoon* were recorded, one in Europe one in Japan and one in India, it is possible that many of the European records refer to different, but at that time unrecognised, species of this genus. Whether this is the case or not, specificity is apparent at the family level, for the hosts in Europe and Asia, the palaearctic zoogeographical region, with the curious exception of the sturgeon, belong to the carp family (Cyprinidae), while the hosts in the neo-tropical and Ethiopian regions are the allied characin family (Characinidae). The type host of *D. paradoxum* is the bream *Abramis brama* (L.).

Diplozoon is a compound animal, formed of two individuals which are so indissolubly united in the region of the apertures of their reproductive glands that their tissues are continuous. It lives attached to the gills of its host, gripping the gill-filaments with its opisthaptors, the two partners sometimes attached to filaments on opposite sides of one gill arch or to filaments on different gill-arches. In this position it can feed, rupturing the tissues of the gill-filament with its powerful prohaptor and imbibing the blood of its host. Sexual maturity is reached in the spring, following a rise in temperature of the water, but it has not been determined whether reproduction is related more directly to that of the host. Oviposition follows, and the eggs (Fig. 15A), each with its shell produced into a single long polar filament by which it is attached to the gill-filaments, hatch in about a fortnight. The larva, a ciliated, free-swimming oncomiracidium (Fig. 15B), has a simple sac-like gut and a pair of closely set eye-spots, and is armed with a pair of comparatively large hooks and a single pair of clamps. It is able to swim actively for a few hours, and if it encounters the appropriate fish in that time it enters the branchial chamber, attaches itself to the gill-filaments with hooks and clamps and loses its cilia. If it is unable to meet with a suitable host it becomes less and less active and dies. The young larva, devoid of cilia, is known as a diporpa (Fig. 15C) and it grows slowly, a second pair of clamps making its appearance in front of the terminal pair. Two curious organs, unique in Platyhelminthes, now make their appearance. They are a button-like projection in the mid-dorsal line situated a little behind the centre of the animal, and a sucker on the ventral surface, also in the mid-line and a little behind the level of the button. The size of the button corresponds to that of the cavity of the sucker and these two bodies may be compared with the two components of a press-stud.

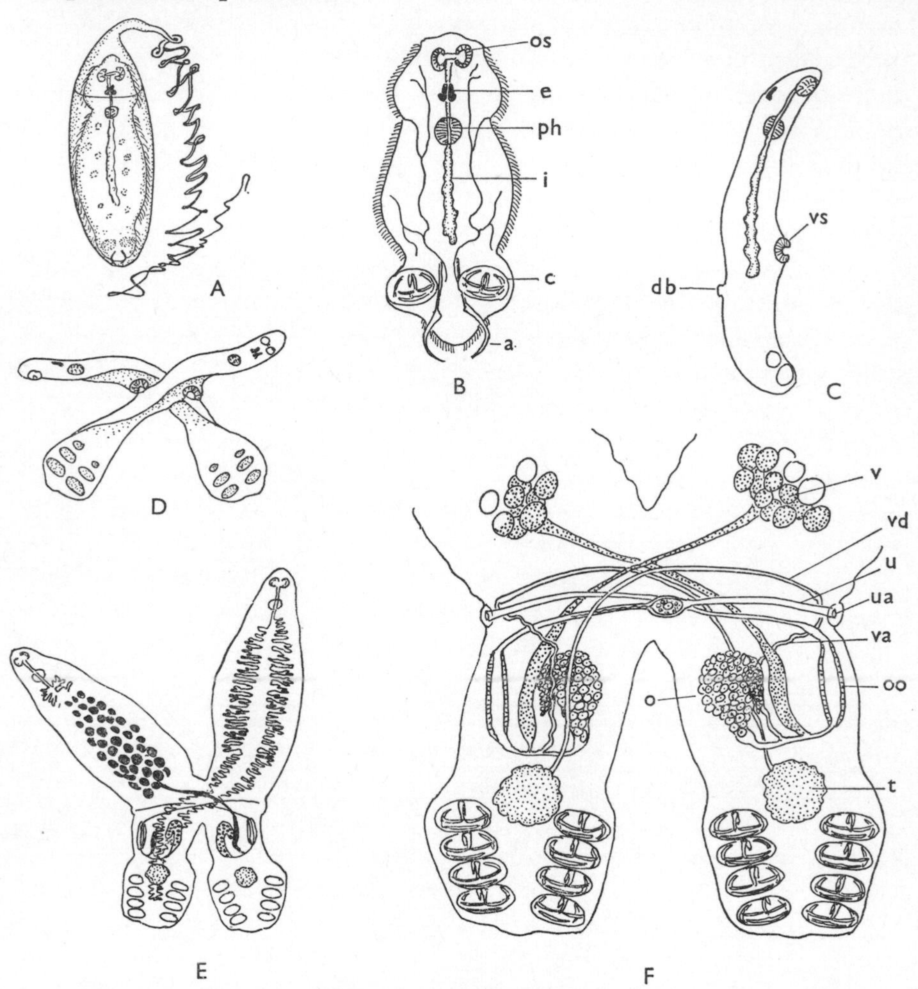

FIGURE 15 *Diplozoon paradoxum* v. Nordmann from the gills of the minnow, *Phoxinus phoxinus* (L.).

A, egg with polar filament and contained embryo; B, free-swimming oncomiracidium that becomes attached to the gills of the minnow; C, diporpa that has not encountered a partner; D, union of two diporpae; E, mature *D. paradoxum*, the gut being omitted from one partner and the vitelline gland from the other; F, reproductive system (semi-diagrammatic). a, larval anchor; c, clamp; db, dorsal button; e, eye; i, intestine; o, ovary; oo, ootype; os, oral suckers; ph, pharynx; t, testis; u, uterus; ua, uterine aperture; v, vitelline gland; va, vagina; vd, vas deferens; vs, ventral sucker.

Larvae move over the gills gripping the filaments alternately with mouth and then with hooks and clamps, and when two are present within one gill-chamber they soon meet. When this encounter takes place one diporpa grips the button of the other with its ventral sucker, and the second diporpa, twisting itself around the first, attaches itself to the first in like manner (Fig. 15D). Once joined the larval tissues unite and the pair takes on the form of a St Andrew's cross, and the clamps and hooks of both participate in attachment to the host. This double animal, which is *Diplozoon*, becomes adult with the development of the third and fourth pairs of clamps in each partner, and the reproductive ducts appear, the sperm-duct of each opening into the vagina of the other (Figs. 15E, F).

Diplozoon lives for a number of years producing eggs in the summer months but showing a regression of the reproductive system in the winter. As a consequence of the extended period of reproduction it can happen that larvae of different ages come together, but, as has been seen, the growth of the larva is slow so their relative sizes are never disproportionate. If a diporpa does not meet with another then it fails to develop and dies before winter.

The site of infestation, the hosts and the geographical distribution of a number of representative species of the different orders of Monogenea are shown in Table 1.

TABLE 1

MONOGENEA—Summary of sites of infection and hosts of representative species

Species	Sites of infestation and hosts	Distribution
CAPSALOIDEA		
Entobdella hippoglossi (Müller)	skin of halibut (*Hippoglossus hippoglossus*), etc.	N. Atlantic, N. Pacific
Capsala martinieri Bosc	skin of sun-fish (*Mola mola*)	N. Atlantic, N. Pacific
Trochopus tubiporus (Diesing)	gills of gurnards and sea-bream	North Sea, Mediterranean
Calicotyle kröyeri Diesing	gills, cloaca, rectum of rays and cloaca of *Chimaera monstrosa*	Eastern N. Atlantic, Mediterranean
Merizocotyle diaphana Cerfontaine	gills of skate (*Raja batis*, *R. oxyrhynchus*)	N. Atlantic
Monocotyle myliobatis Taschenberg	eagle-ray (*Myliobatis aquila*)	Mediterranean
Dictyocotyle coeliaca Nybelin	coelom of rays	N. Atlantic, North Sea
Dactylogyrus vastator Nybelin	gills of carp (*Cyprinus*, *Carassius*)	Fresh water Europe, Asia
Tetraonchus monenteron (Wagener)	gills of pike (*Esox lucius*, *E. reicherti*)	Fresh water Europe, N. America
Acolpenteron ureteroecetes Fischthal et Allison	ureters, urinary bladder, black bass (*Micropterus salmoides*), small-mouth black bass (*M. dolomieui*)	Fresh water central N. America
Tristoma coccineum Cuvier	gills of swordfish (*Xiphias gladius*), hammerhead shark (*Sphyrna zygaena*)	Indian Ocean

Table 1 continued overleaf

TABLE 1—*continued*

Species	Sites of infestation and hosts	Distribution
GYRODACTYLOIDEA		
Gyrodactylus elegans v. Nordmann	gills of many fresh-water, and some marine, teleosts	Europe, N. America
ACANTHOCOTYLOIDEA		
Acanthocotyle Monticelli (3 spp.)	skin of thornback (*Raja clavata*)	Eastern N. Atlantic, Mediterranean
PROTOGYRODACTYLOIDEA		
Protogyrodactylus quadratus Johnston et Tiegs	gills of fresh-water fishes (*Therapon* sp.)	Queensland
UDONELLOIDEA		
Udonella caligorum Johnston	on marine, parasitic copepods (*Caligus* sp.) and their marine hosts (*Hippoglossus*, *Gadus*, *Trigla*, etc.)	cosmopolitan
CHIMAERICOLOIDEA		
Chimaericola leptogaster (Leuckart)	gills of *Chimaera monstrosa*	N. Atlantic
DICLYDOPHEROIDEA		
Diclidophora denticulata (Olsson)	gills of some gadoids, coalfish (*Gadus virens*) and hake (*Merluccius merluccius*)	N. Atlantic
Diplozoon paradoxum Nordmann	gills of many fresh-water fishes, e.g. carp, burbot, roach, rudd, minnow, etc.	Europe, Asia, India, N. America
Cyclocotyla bellones Otto	dorsal skin of garfish (*Belone bellone*)	Eastern N. Atlantic, Mediterranean
Anthocotyle merlucci v. Beneden et Hesse	gills of hake (*Merluccius merluccius*)	North Sea
Mazocraës alosae Hermann	gills of shad (*Alosa brashnikovi*) and other *Alosa* spp.	European waters
Kuhnia scombri (Kuhn)	gills of mackerel (*Scomber scombrus*) and other *Scomber* spp.	N. Atlantic, N. Pacific
Microcotyle donavini v. Beneden et Hesse	gills of ballan wrasse (*Labrus bergylta*)	Eastern N. Atlantic
POLYSTOMATOIDEA		
Polystoma integerrimum (Fröhlich)	urinary bladder of certain frogs and toads	Europe, N. America (Quebec), Africa (Congo)
Oculotrema hippopotami Stunkard	eye of hippopotamus	Egypt, Uganda
Sphyranura osleri Wright	skin of mud-puppy (*Necturus maculatus*)	N. America
DICLYBOTHRIOIDEA		
Rajonococotyle batis Cerfontaine	gills of *Raja batis*, *R. binoculata*, *R. oxyrhynchus*	N. Atlantic, N. Pacific
Diclybothrium armatum (Leuckart)	gills of Chondrostei—(*Acipenser* sp.)	Europe, N. America

6 *The Aspidogastrea*

'Curiouser and curiouser' said Alice.
LEWIS CARROLL

This small group of parasites, amounting in all to about thirty described species, has been seen (Chapter 2) to differ as much from monogeneans as from digeneans. To recapitulate, the characters that separate them from these two classes are (*a*) the sole-like adhesive organ, (*b*) an unbranched gut, (*c*) peculiar marginal organs and (*d*) the absence of a metamorphosis. To these one might add the restriction to certain kinds of hosts, marine crustacea, marine and fresh-water molluscs and marine and fresh-water fishes and chelonians.

ASPIDOGASTREAN HOSTS

An analysis of aspidogastreans with regard to the kind of hosts in which they are found shows that maturity in some of them may not be obtained in an invertebrate host (mollusc or crustacean) but in a vertebrate host (fish or chelonian) that preys on it. A second group comprises those that become sexually mature in the first host, a mollusc, and may or may not reach a second host which can accordingly be regarded as a facultative host. The third group becomes sexually mature directly in a vertebrate host. Because the two hosts of some forms are a mollusc and a vertebrate one naturally recalls the obligatory, molluscan, intermediate host of digeneans where the definitive host is a vertebrate and looks for evidence in support of neotenic or progenetic development. This possibility has been dismissed by Dollfus who insists that there is nothing in their morphology nor their anatomy to suggest that this is the case and that neither those attaining maturity in a mollusc nor those requiring two hosts can be regarded as the more primitive.

ASPIDOGASTER CONCHIOLA

The first to be described, and the most widely known member of the class is *A. conchiola* v. Baer which is a common parasite in fresh-water mussels (Unionidae) on both sides of the Atlantic. It is found as a mature, egg-producing form in the pericardium or those parts of the kidney known as Keber's organ and the organ of Bojanus which are associated with the pericardium, and the eggs reach the exterior through the excretory duct. The young, on hatching, show a single postero-ventral undivided sucker and an oral one and they lack cilia (Fig. 16). It is not known how they reach the peri-

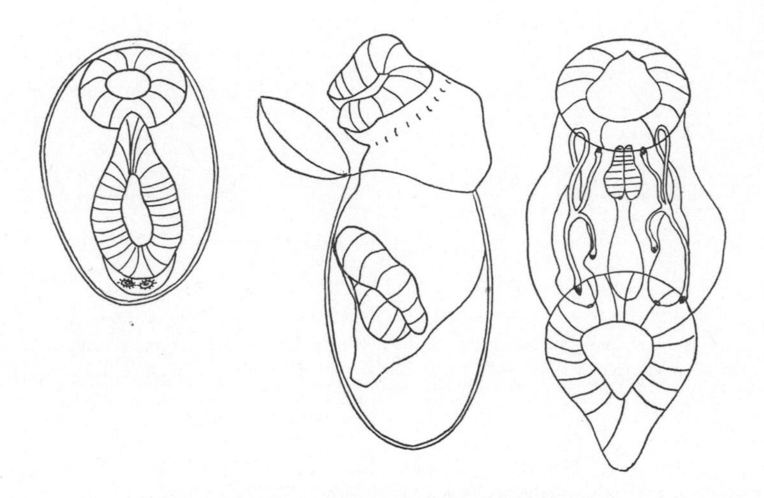

FIGURE 16 *Aspidogaster conchiola* v. Baer.
Egg with contained embryo, larva emerging, and larva showing oral and ventral suckers and excretory system. (After Dollfus from Faust.) × 350.

cardium of the host but once there development is slow, the larval sucker being gradually replaced by the alveoli that form the adult adhesive organ. The frequency of the occurrence of *A. conchiola* (Fig. 6) in mussels leaves no doubt they are the normal hosts while the occasional records of its occurrence in the gut of the fresh-water chelonian *Amyda sinensis* and a dace (*Leuciscus aethiops* Basilewsky) in the Far East, and attached to the wall of the intestine in the carp (*Cyprinus carpio* L.) and the redhorse (*Moxostoma macrolepidotum* Le Sueur) in North America indicate its ability to tolerate other hosts. This has been shown experimentally by van Cleave and Williams who injected living, mature *A. conchiola* into the gut of a water-tortoise *Pseudemys troosti* (Holbrook) and found one of them still alive and attached to the wall of the stomach a fortnight later.

COTYLOGASTEROIDES OCCIDENTALIS

Cotylogasteroides occidentalis (Nickerson), a species allied to the above, is found in the prosobranch mollusc *Goniobasis* sp. in Lake Erie and also in the mussel *Lampsilis luteolus* (Lamarck). A vertebrate host of this species is the North American fish known as the sheepshead, a sciaenoid fish (*Aplodinotus grunniens* Rafinesque). Feeding experiments, in which five sheepsheads were forcibly fed with large specimens of *C. occidentalis* from *Goniobasis*, one each day for five days, and on examination five days later all the worms were recovered from the hind gut, show that *C. occidentalis* is also a facultative parasite.

The development of *C. occidentalis* differs somewhat from that of *A. conchiola* because the larva, called a *cotylocidium*, possesses two discontinuous bands of tufts of cilia, one near the front and one near the hind end and is free-swimming. Less primitive, however, is the larva of an aspidogastrean found in the loggerhead turtle *Thalassochelys caretta* (L.). This is *Lophotaspis vallei* Stossich, the genus differing from *Aspidogaster* in the possession of numerous papillae on the ventral surface and with tube-feet-like marginal organs. The larva swims by means of three patches of cilia, but with a rudimentary adhesive organ behind and an oral sucker in front it can also creep like a leech. One of the records of this aspidogastrean is from the stomach and oesophagus of loggerhead turtles taken in the Indian Ocean near Ceylon and this region is where *L. margaritiferae* (Shipley et Hornell) is found in the pericardial cavity of the pearl oyster, *Margaritifera vulgaris* Schumacher. The specimens of *L. margaritiferae* were not adult forms, but these two aspidogastreans are otherwise so alike that it is difficult to imagine that they are not one and the same species, and that *L. margaritiferae* is an immature form that attains maturity in the turtle.

STICHOCOTYLE NEPHROPIS

Stichocotyle Cunningham is an aspidogastrean that stands alone, differing in a number of respects from all the others. In its structure its adhesive apparatus is formed of a single, longitudinal row of independent suckers (Fig. 7) and not a single, compound organ and it lacks the marginal organs characterising the other genera. Another fundamental difference is that a larval or nymphal encysted stage is found in the wall of the intestine of a decapod crustacean while the adult stage is in the bile-duct of an elasmobranch fish in which it may attain a length of five inches, appearing as a long slender worm with as many as 30 suckers in a row along its ventral surface. *Stichocotyle nephropis* Cunningham, the only species so far described, is found in the Norway lobster (*Nephrops norvegicus* L.) in the Firth of Forth, and in the American lobster (*Homarus americanus* M.-Edwards) on the other side of the Atlantic. Mature sexual forms, however, live in the predators on these decapods, in the bile-duct of the thornback ray (*Raja clavata* L.) in the North Sea, and the barndoor ray (*Raja laevis* Mitchill) of the western North Atlantic.

MACRASPIS ELEGANS

One can round off the picture of the Class Aspidogastrea with *Macraspis elegans* Olsson (Fig. 17), a parasite that lives in the gall-bladder of chimaeras. As the chimaeras (Holocephali) have a very long ancestry reaching back to Devonian (Old Red Sandstone) or Carboniferous times, since when they have existed as a separate group with the sharks as their nearest living relatives, it is not unexpected to find that they have distinctive parasites. Amongst these are the aspidogastrean *Macraspis* and the cestodarian *Gyrocotyle*, the latter of which will be discussed later. Chimaeras are found in the colder waters of the northern and southern hemispheres, and the two species *Chimaera monstrosa* L., commonly called the rabbit-fish or the king of the herrings of the North Atlantic, and *Callorhynchus milii* B. de St V., in the seas of New Zealand, are both hosts of *M. elegans*. Other species of this genus sometimes referred to as a separate sub-genus are recorded from a number of sharks and dogfishes, and an immature form in the teleostean sciaenid, *Menticirrus americanus* (L.) in the intervening warmer waters.

The most conspicuous region of the body of *M. elegans* is the adhesive muscular sole which is worth more detailed consideration. It extends under the greater part of the body (Fig. 17) with only a short, cylindrical, anterior neck terminating in a conspicuous sucker projecting in front of it. The sole is a compound structure, appearing to be divided by transverse septa into as many as 100 alveoli. In young forms the alveoli are few, but new alveoli appear at the posterior end of the organ where, from a blastema, myoblasts become organised to form the muscles of each additional part of the sole. Thus the posterior border of the last-formed alveolus becomes modified to become a transverse septum as a new alveolus is formed behind it, so that the sole is not a single subdivided sucker but a series of closely set suckers whose contiguous borders are transformed into transverse septa. This can be deduced from the arrangement of the muscles in the sole. There are radially arranged fibres in each alveolus, the contraction of which will cause their walls to become thinner with corresponding enlargement of their cavities. Antagonistic muscles are seen in the layers of transverse fibres that lie immediately below the cuticle, while internal to these, extending along the whole length of the sole, are the longitudinal fibres. This arrangement is such that at the lateral margins of the transverse septa there is a region, triangular in horizontal section, from which these three main layers are wanting. It is here that the peculiar marginal organs are found, which are manifest at the surface as lateral papillae. Each papilla is formed of a cylindrical column surmounted by a disc-like apex, in the centre of which is a depression leading into a funnel-like aperture internal to the narrow end of which lies a

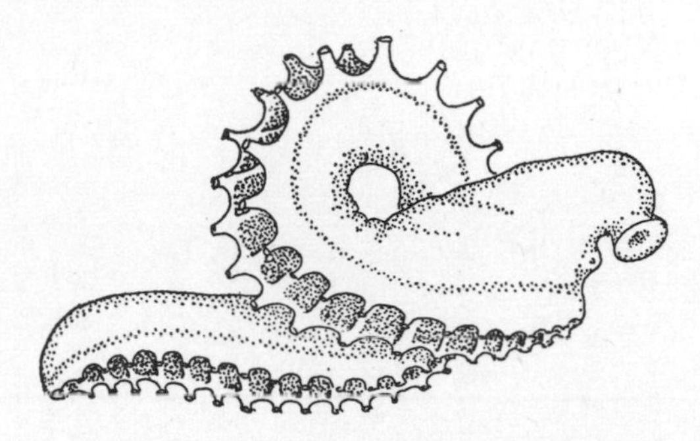

FIGURE 17 *Macraspis elegans* Olsson, from the gall-bladder of *Chimaera monstrosa* L. (up to 22 mm long).

mass of peculiar cells which constitute the marginal organs. Nerve fibres ramify through these cells and they are connected with the lateral nerves of the body through the nerve-net that lies over the dorsal surface of the adhesive sole. Some authorities (e.g. Dollfus) attribute a secretory function to them, while others (e.g. Baer) consider them to be sense organs, but their true function remains to be elucidated.

Little is known regarding the development of *M. elegans*. Embryonated eggs from the uterus have embryos similar to those of *Aspidogaster* (Fig. 16) and measure about 0·1 mm in length, while the youngest forms known from the bile-ducts of *Chimaera* are quite immature and are 1·5 mm long, but nothing is known of the intervening stages, though these facts suggest that there is but one host.

A summary of the hosts, sites of infestation and distribution of the aspidogastreans discussed in this chapter is seen in Table 2.

TABLE 2

ASPIDOGASTREA—summary of representative species

Species	Host	Site	Distribution
Aspidogaster conchiola v. Baer	lamellibranchs (Unionidae, etc.)	adult, in pericardium, kidney	Europe, N. America
	reptile (*Amyda sinensis*)	adult, in intestine	China
	fish (*Leuciscus aethiops*)	adult, in intestine	China
	gastropod (*Vivipara* spp.)	adult, in visceral mass	Egypt, China
Cotylaspis insignis Leidy	lamellibranchs (*Anodonta* spp.)	adult, in kidney, pallial cavity	N. America
Lophotaspis vallei (Stossich)	turtle (*Thalassochelys caretta*)	adult, in oesophagus, stomach	Atlantic, Mediterranean
	gastropod (*Fasciolaria gigas*)	juv., in pallial cavity	Florida
=*Lophotaspis margaritiferae* (Ship. et Horn.) (probably)	oyster (*Margaritifera vulgaris*)	juv., in pericardium	Indian Ocean—Cheval Paar
Cotylogasteroides occidentalis (Nickerson)	lamellibranch (*Unio* sp.)	juv., in branchial cavity	N. America
	prosobranch (*Goniobasis* sp.)	adult	N. America
	fish (sheepshead—*Aplodinotus grunniens*)	adult, in intestine	N. America
Stichocotyle nephropis Cunningham	lobster (*Nephrops norvegicus*)	encysted in wall of intestine	Eastern N. Atlantic
	thornback ray (*Raja clavata*)	adult, in bile-ducts	Eastern N. Atlantic
	lobster (*Homarus americanus*)	encysted in wall of intestine	Western N. Atlantic
	barndoor ray (*Raja laevis*)	adult, in bile-ducts	Western N. Atlantic
Macraspis elegans (Olsson)	Holocephali (*Chimaera monstrosa*)	adult, in gall-bladder	N. Atlantic
	Holocephali (*Callorhynchus milii*)	adult, in gall-bladder	S. Pacific

7 *The Digenea, Trematoda or Flukes*

But Facts are chiels that winna ding,
An' downa be disputed.
ROBERT BURNS

The Digenea or trematodes, popularly called flukes, are in their adult state found as obligate endoparasites in all classes of vertebrates. They are usually flattened animals, varying in size from a fraction of a millimetre to several centimetres in length, and provided with at least one organ of adhesion in the form of a sucker.

Their life-cycles, which usually involve one or two free-living stages, show variations on a theme which is fundamental to all in this class. The theme is as follows: the adult, parasitic in a vertebrate, produces eggs from which hatch miracidia; the miracidia develop further in a mollusc in which various reproducing forms succeed one another, and from the mollusc there emerge eventually free-swimming larvae known as cercariae which gain access to the definitive host actively or passively. There may be a second or even a third intermediate host in which the cercaria is encysted (Fig. 18).

PARORCHIS ACANTHUS

This fluke has already been discussed as an endoparasite that has to contend with a number of different environments during its life-cycle (Fig. 19), and it is considered here as a typical member of its class, and this is taken as a basis for a discussion of the origin, evolution and adaptations of these animals. It is most commonly found in the herring gull, though it is found in other birds, and its habitat is the *bursa Fabricii*, or the rectum. It is hermaphrodite but it is unlikely that it is self-fertilised. The eggs within the uterus are seen in various degrees of development, from unsegmented eggs, lying close to the ovary, through developing embryos to well-developed embryos, ciliated and complete with eye-spots and penetration glands, and finally active miracidia which have hatched within the uterus. Within each miracidium is a developing redia which will become the next stage in the life-cycle, and within the redia are cells, dividing to form little masses of cells called germinal balls, which will eventually develop into daughter rediae. The miracidia leave the

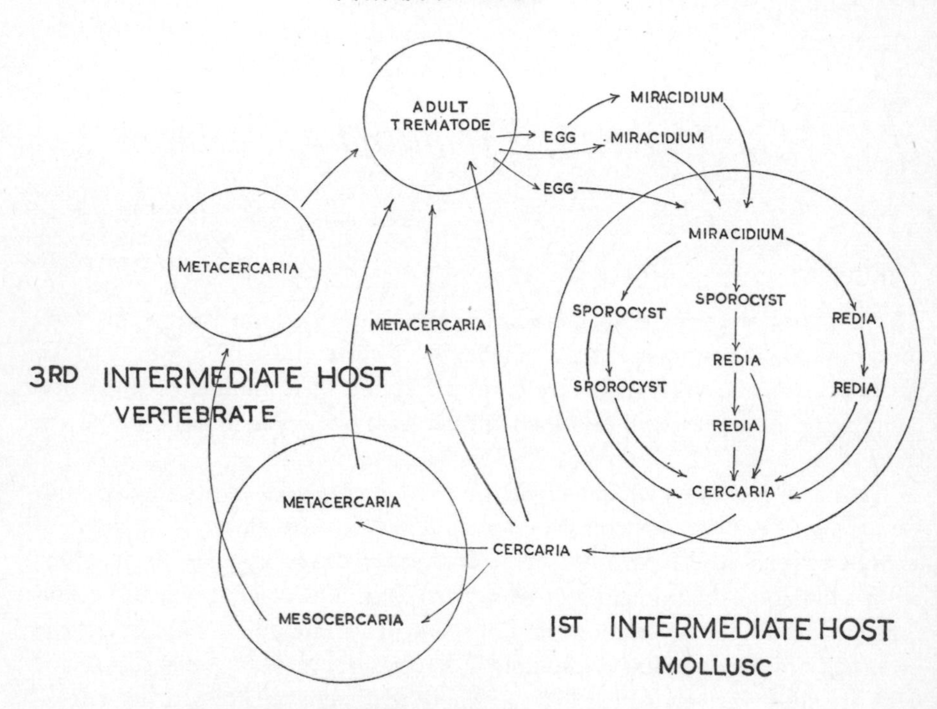

FIGURE 18 Scheme showing variation in the life-cycles of digenetic trematodes. (After J. G. Baer.)

fluke through its genital pore and pass out of the rectum in the droppings of the gull. If these fall into a rock-pool the miracidia then have the opportunity of infesting the whelk, *Nucella lapillus* (L.). The miracidium attaches itself to the whelk and, assisted by the secretions of its penetration glands, bores into the whelk's body. The redia is liberated, and the germinal balls within it develop into some 20 daughter rediae, which later escape through the birth pore behind the collar, and tear their way through the tissues to the surface of the hepato-pancreas of the whelk, using as organs of locomotion the muscular

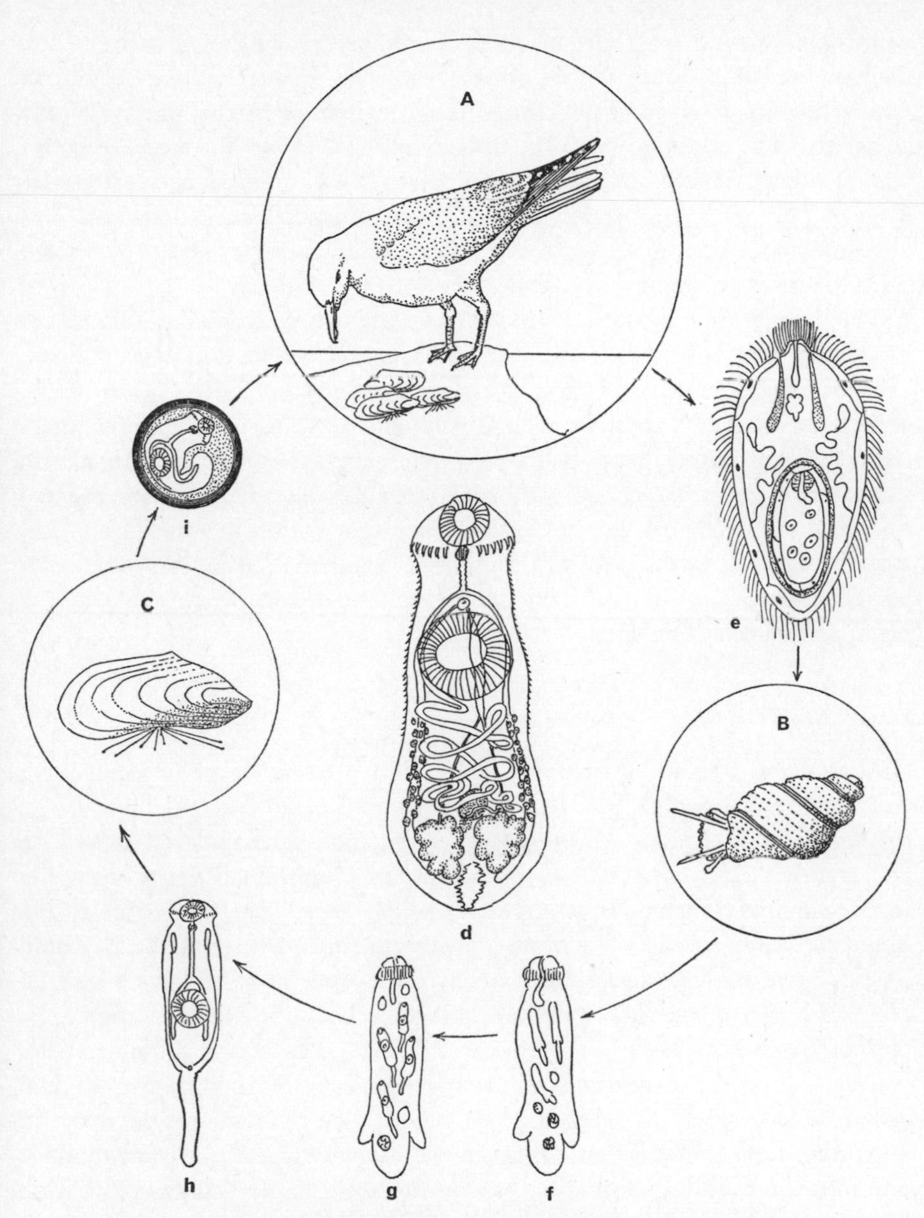

FIGURE 19 *Parorchis acanthus* Nicoll.
A, definitive host, the herring gull, *Larus cinereus* Brisson; B, first intermediate host, the whelk, *Nucella lapillus* (L.); C, second intermediate host, the mussel, *Mytilus edulis* L.; d, adult fluke from *bursa Fabricii* of herring gull; e, miracidium from uterus of *Parorchis acanthus* containing developing redia; f, redia containing developing daughter rediae; g, daughter redia containing cercariae; h, cercaria; i, metacercaria from tissues of mussel.

collar in front and their lateral processes or procruscula. In this site the daughter rediae produce within themselves, in a manner similar to that in which they themselves were formed, the next stage in the life-cycle, the cercariae. The cercariae leave the rediae, make their way through the body-wall of the whelk and once again there is a free-swimming stage in the life-cycle. The cercariae penetrate cockles or mussels, lose their tails and secrete a cyst, being now known as metacercariae, and encysted they remain until the mollusc is eaten by a gull.

The phenomenon of reproducing forms which succeed one another in the molluscan host, viz. miracidium, redia, daughter rediae and cercariae has intrigued many, and there have been various interpretations. One of the earliest was that of Steenstrup who looked on it as being similar to the alternation of generations met with in Zoophytes, such as is seen in *Obelia* or in the parasitic *Polypodium*, where a sexual generation alternates with an asexual one, in this case the sexual generation being that within the vertebrate host and the asexual one that in the mollusc. Another explanation labelled it parthenogenesis, or virgin birth, but it is more reasonable to consider the phenomenon as a form of *polyembryony*.

POLYEMBRYONY

As the name implies, polyembryony is the production of a number of embryos from a single ovum and is encountered in different animals. The phenomenon is probably most easily understood if considered where its occurrence is less complicated as for instance in certain small wasps belonging to the families Braconidae or Ichneumonidae which lay their eggs in the bodies of other insects. Such an egg, as for example that of the braconid *Macrocentrus homonae*, is deposited within the caterpillar of the tea-tortrix, *Homona coffearia*, and then it divides to form a chain of a dozen or more eggs, each of which develops into a larva. In this case all the larvae produced from one egg are at the same stage of development, the division of the original ovum to form many ova taking place before development into larvae. If, however, one can imagine a time-lag in the development of the ova produced one sees the condition obtaining in the monogenean *Gyrodactylus*. Within the body of *Gyrodactylus* one can see an embryo which has developed from a fertilised egg, and within this embryo is a second embryo at an earlier stage of development, and within the second embryo, at a still earlier stage of development, is yet another, and finally the beginning of a fourth embryo within the body of the third. In this example all the developmental forms are similar, which is not, however, the case in Digenea.

In *Parorchis* the fertilised ovum is the starting-point; it gives rise to a miracidium and a contained embryonic or germinal cell. This latter cell divides, not only to form the redia that develops within the miracidium but also to produce germinal cells which develop later within it into daughter rediae. The daughter rediae in turn develop and form within themselves in a similar manner, the cercariae. There is thus a direct continuity of germinal or embryonic cells, giving rise in turn to the different larval stages, miracidia, redia, daughter rediae and cercariae.

STAGES IN THE LIFE-CYCLE

Various species of flukes exhibit differences which may involve every stage in the life-cycle. The more usual variations are seen in Fig. 18. The host of the adult fluke is a vertebrate animal in which a particular organ comes to be occupied as the normal site of infestation. Some part of the alimentary canal is the most common site. *Paramphistomum* Lühe and *Gastrothylax* Poirier (Fig. 20) are found in the rumen of cattle; *Echinostoma* Rud. in the intestine of birds; *Fasciola* L. and *Dicrocoelium* Duj. in the bile-ducts of sheep; *Clonorchis sinensis* Cobbold (Plate 1 and Fig. 21) in the bile-ducts of cat,

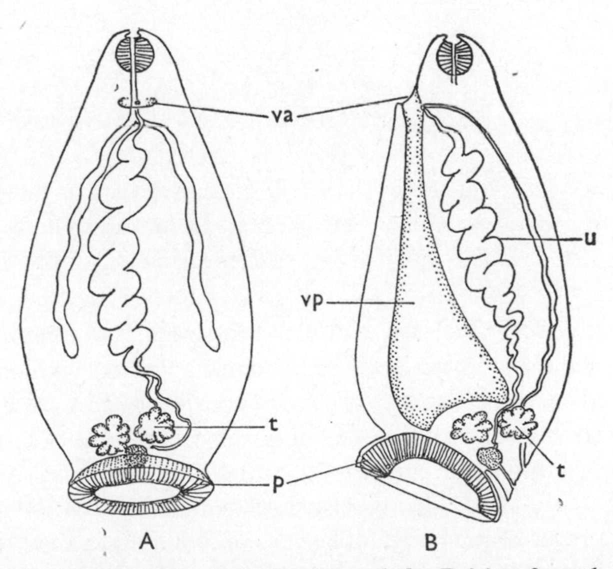

FIGURE 20 A paramphistome fluke, *Gastrothylax* Poirier, from the rumen of the zebu, *Bos indicus* L., showing the large ventral pouch which opens close to the oral sucker.
A, ventral view, B, side view. p, posterior sucker; t, testis; u, uterus; va, aperture of ventral pouch; vp, ventral pouch.

dog and man; while the intestinal caeca and *bursa Fabricii* of birds are also sites. In the respiratory system there are lung-flukes such as *Haplometra* Looss in the lungs of frogs, *Cyclocoelum* Brandes in the air-sacs and nasal sinuses of birds, and others in the tracheae and bronchi. The excretory system as a niche is not excluded, for among kidney-flukes is *Renicola* Cohn (Fig. 8) while *Gorgodera* Looss is often met with in the bladder of frogs. From such sites there is always passage available whereby eggs can reach the

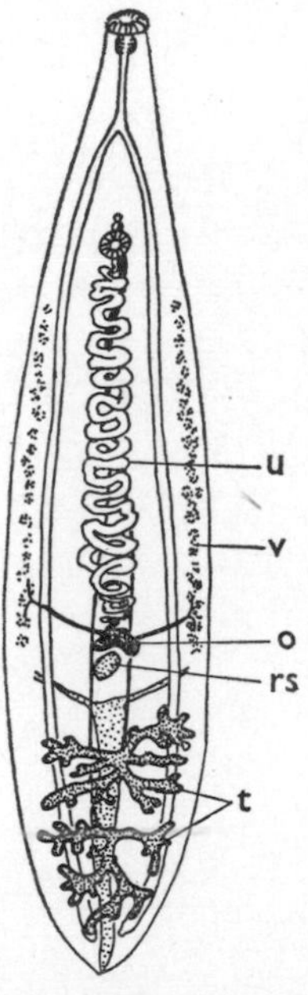

FIGURE 21 *Clonorchis sinensis* Cobbold from the liver of man, cat or dog. o, ovary; rs, receptaculum seminis; t, testes; u, uterus; v, vitelline glands.

exterior but this is not the case with blood-flukes such as *Schistosoma* Weinland of man and other animals where the adult flukes occupy veinules of the mesentery or bladder. The egg of *Schistosoma* is armed with a sharp spine and when oviposition takes place within a veinule of some organ such as the intestine or the bladder the contraction of the organ concerned causes the egg to penetrate the wall of the blood-vessel and to be forced mechanically through the tissues of the body. The eggs of *Schistosoma bovis* (Sonsino) are elongate and spindle-shaped, produced into a sharp process at one end (Fig. 22d) while those of *S. mansoni* Sambon, *S. haematobium* (Bilharz) and *S. japonicum* Katsurada are ovoid with a sharp lateral or terminal spine serving the same function.

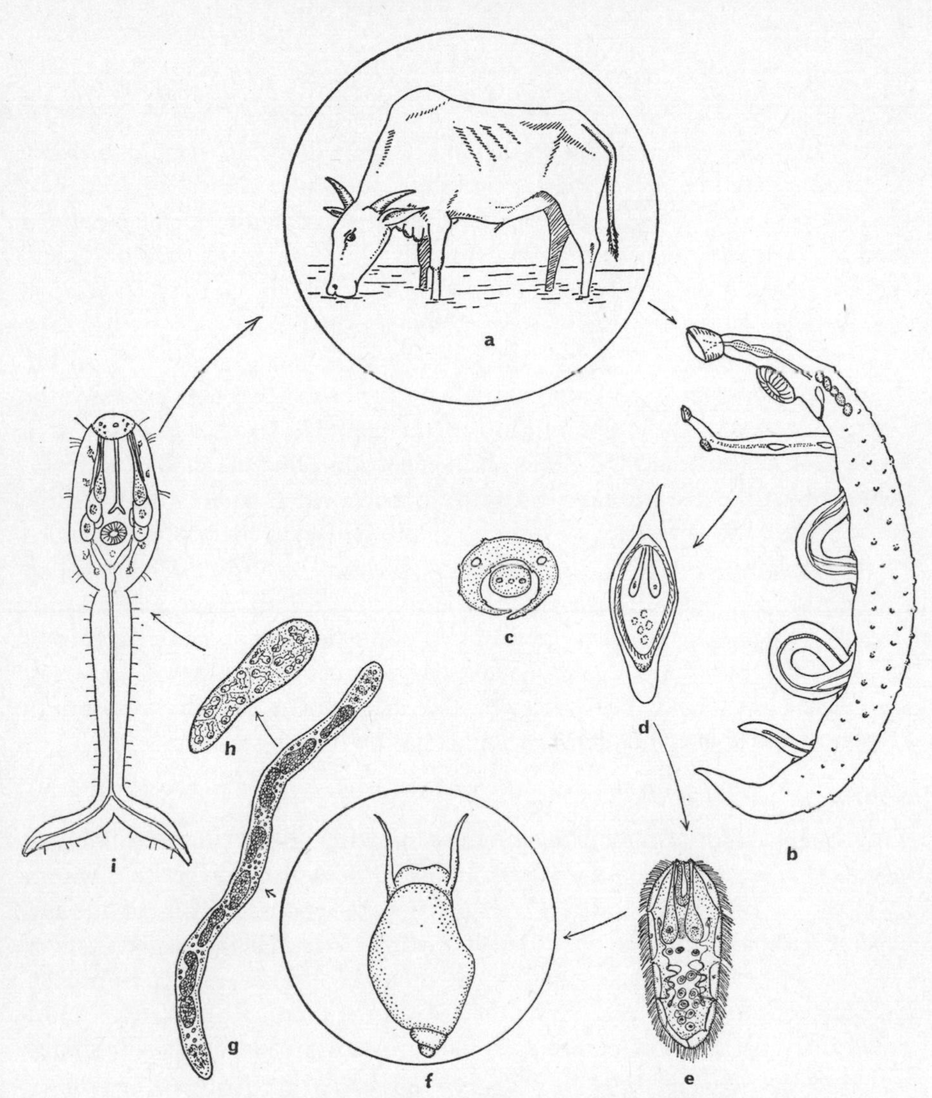

FIGURE 22 Life-cycle of *Schistosoma bovis* (Sonsino).
a, the definitive host, a cattle-beast such as the zebu, *Bos indicus* L.; b, male and female *S. bovis* removed from a mesenteric vein of the definitive host; c, section of the schistosomes showing the female lying in the gynecophoric groove of the male; d, egg with contained miracidium; e, free-swimming miracidium showing penetration glands; f, the intermediate molluscan host *Bulinus* sp. which the miracidium penetrates and in which it develops into a sporocyst (g) containing daughter sporocysts (h). The furcocercaria (i) is liberated from the daughter sporocyst and the mollusc and penetrates the definitive host through the skin or mucous membrane.

Miracidium

The next stage in the life-cycle is the miracidium, which at first sight looks like a ciliated protozoon, and indeed, when first recorded, was described as such. It is free-swimming with a body contained within some two dozen flattened ciliated cells. A short, blind gut opens on a conical projection at the anterior end and one or more pairs of penetration glands open close to the mouth. One or more pairs of protonephridia are present with their ducts and there is a cluster of germ cells which eventually gives rise to the next stage in the life-cycle. Many miracidia have a pair of eye-spots.

As seen in *Parorchis* (Fig. 19), the miracidium is hatched while still within the uterus of the adult. A more common occurrence is that the eggs reach the exterior and hatch under the influence of temperature, light or moisture as in *Fasciola* or *Schistosoma*. In either case, on encountering the molluscan intermediate host the miracidium penetrates its body either directly through the body wall or less directly by entering the pulmonary chamber first, enabled by the enzymes secreted by its penetration glands. A third possibility is that the eggs do not hatch until eaten by the molluscan host, as in *Dicrocoelium* (Fig. 23) and in this case active penetration of the host is not necessary though of course the miracidium has to make its way out of the gut into the tissues of the hepato-pancreas. But whatever the mechanism the end result is a constant feature of the Digenea, a miracidium within a mollusc.

Sporocyst and Redia

The possible fates of the miracidia are seen in Fig. 18. A miracidium may lose its cilia and become a sporocyst, a hollow sac with germinal cells within its cavity. These cells, of direct lineage from the fertilised ovum, divide and become organised into the next stage in the life-cycle, daughter sporocysts or rediae. The redia is more specialised than the sporocyst. It is usually elongate with a conical anterior end on which the mouth opens from a short blind gut. Behind this region there is a muscular collar and in the mid-region of the body or towards the posterior end there is a pair of hollow, lateral processes, the procruscula, which are directed posteriorly. Within the redia there are flame-cells and their ducts and in addition the germ-cells derived directly from those of the miracidium or sporocyst. The organs used in locomotion are the muscular collar and the procruscula associated with the muscular fibres in the body wall which enable the body to contract or extend. The redia moves through the tissues of the snail by extending its body when the procruscula give some purchase and enable the anterior end to be pushed forwards. When the muscular collar expands it grips the tissues and the hinder part of the body is drawn forwards and then the whole process is repeated. In this way the redia reaches the hepato-pancreas of the snail, the usual site of infestation, with its readily available supply of food.

In the case of *Fasciola* each sporocyst produces from five to eight rediae, and these move to the hepato-pancreas where each redia produces within itself some eight to twelve rediae of the second generation. In *Dicrocoelium* and *Schistosoma* there may be two generations of sporocysts, the daughter sporocysts developing within the first and on being liberated making their way, though precisely how they do this is not known, to the hepato-pancreas. The miracidium of *Parorchis*, as we have seen, gives rise directly to a redia which is formed within it before hatching.

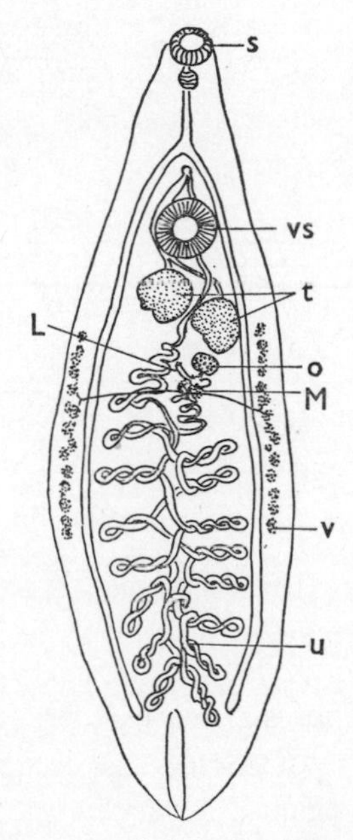

FIGURE 23 The lesser liver-fluke of sheep, *Dicrocoelium lanceatum* (Rudolphi). A comparatively young individual in which the uterus is not fully developed and is diagrammatised to show the twisting of its lateral loops which gives the appearance of a branched uterus in mature forms. L, Laurer's canal; M, Mehlis's gland; o, ovary; s, oral sucker; t, testes; u, uterus; v, vitelline glands; vs, ventral sucker.

Cercaria

Just as rediae were formed within rediae or within sporocysts from germ-cells, in the same way are the cercariae formed. When they are fully developed they make their way out of the sporocyst or redia as the case may be. In the former case the elastic wall closes the puncture in the sporocyst-wall after the cercaria has escaped, though the redia is provided with a birth-pore situated just behind the collar, and through this the cercariae leave.

In its most common form a cercaria has a small body in which one can see something of the characters of the adult as in the relative position of the suckers or the presence of spines, and it usually has a tail which acts as an organ of locomotion. In addition one can see the pattern of the flame-cells and the nature of the excretory bladder. Larval organs may be present. In some (*Parorchis* and *Fasciola*) there are voluminous cystogenous glands which function later in producing a protective cyst around the cercaria, and in others (*Dicrocoelium*) there is a stylet which projects from the mouth and also functions as a penetrating organ. Other larval organs which do not appear in the adult are photoreceptors in the form of eye-spots (*Clonorchis* (Fig. 21)) which may be simple or compound and which sometimes show rudimentary lenses. The cercarial tail also shows certain characteristic features. It may be small or even absent. It may be narrower than the body when the body is provided with a collar of spines (*Echinostoma*, *Parorchis*) or naked (*Fasciola*); the tail may be forked (*Schistosoma* and *Alaria* Schrank) or provided with fin-folds or with spines. In some the tail is large with a cavity in its base into which the body of the cercaria is retracted (*Gorgoderina*). Cercariae can swim and in the case of the cercariae of schistosomes, furcocercariae, in which the tail is forked, swimming takes place with the tail in front. Cercariae can also crawl, leech-like, along the substratum, extending and shortening the body.

VARIATIONS IN THE LIFE-CYCLE

The most direct method of infestation is where, as in *Schistosoma*, the cercaria attaches itself in water to the skin of the definitive host using its adhesive glands, and with its cytolytic glands it penetrates the skin, dropping its tail in the process. Once inside the host's body and having entered a blood-vessel or lymphatic the now tailless cercaria is carried to the liver, in the veins of which it reaches maturity. Infestation may also be brought about by the definitive host swallowing an encysted cercaria, known as a metacercaria. In the case of *Fasciola* they are formed on grass and may be swallowed by grazing sheep. Encystation may take place in another animal, as for example the metacercaria of *Parorchis* within a mussel, or that of *Clonorchis* under the scales of goldfish. In the latter cases the mussel and the goldfish comprise the second intermediate host. Another example of a second intermediate host is that in the life-cycle of *Dicrocoelium* where cercariae are liberated from the snail in one or two slime-balls, which are more or less spherical masses of mucus formed by the snail and the cercariae. Each slime-ball contains some eighty to a hundred cercariae. These may be eaten by a grazing sheep and infestation brought about in this way or they may be eaten by an ant. Within an ant the cercariae penetrate the gut, reach the haemocoel whence they may reach the ant's head. It seems almost incredible that the presence of a cercaria in the head of an ant should affect its instinct in such a way that the ant is irresistibly impelled to climb. Such, however, appears to be the case, and an ant perched at the tip of a blade of grass is more likely to be eaten by a grazing sheep, which then becomes infested.

One may take as an example of three intermediate hosts, the fluke *Alaria mustelae* Bosma, in which as Bosma has shown there are three obligatory

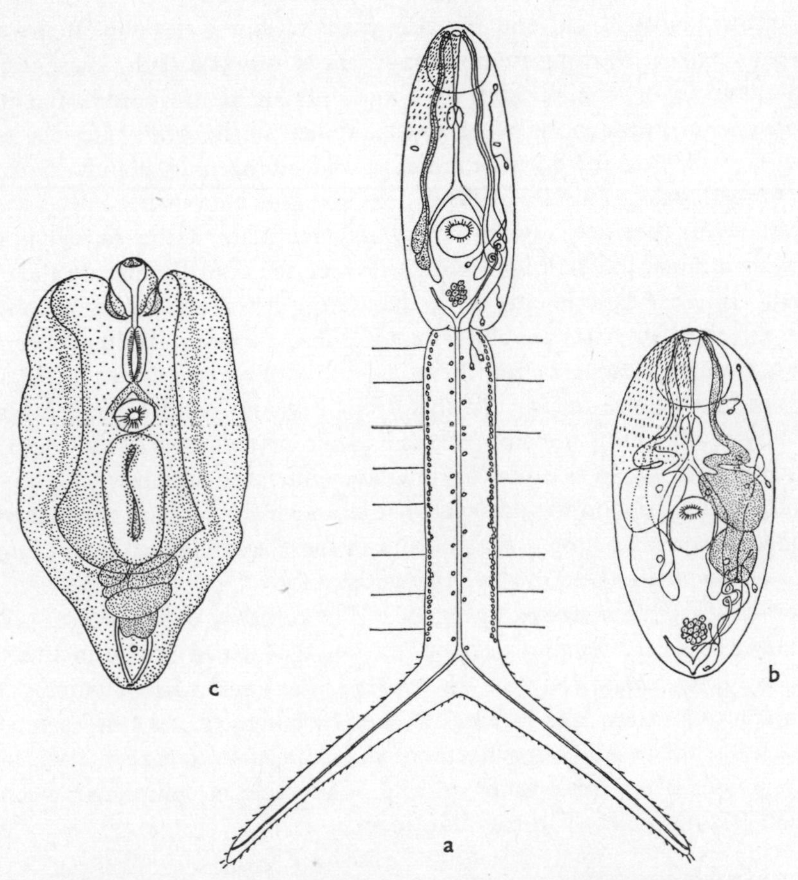

FIGURE 24 *Alaria mustelae* Bosma. Stages from intermediate hosts. a, Furcocercaria, semi-diagrammatic, from *Planorbula armigera*; b, Mesocercaria from a frog; c, Mature metacercaria from a rodent. (After Bosma.)

intermediate hosts, the adult or definitive host being the North American mink *Mustela vison* Schreber. It is a small fluke with an average length of a millimetre and a quarter, and it shows two regions of the body, a scoop-like anterior moiety and a cylindrical hind body. The scoop-like region with its foliaceous lateral expansions and a relatively large, very contractile and protrusible hold-fast organ lying immediately posterior to the ventral sucker enable the fluke to cling closely to the intestinal villi. The first intermediate host, an aquatic snail *Planorbula armigera* (Say) is entered by the miracidium which gives rise to two successive generations of sporocysts. From the second of these, furcocercariae emerge (Fig. 24) and leaving the molluscan host enter frogs or their tadpoles, various species of which can be parasitised—bull-frogs, leopard-frogs, pickerel-frogs or green frogs. In the amphibian the cercaria loses its tail and remains encysted but it develops further, the excretory system showing more solenocytes: it is now called a mesocercaria. If the tadpole, or frog, is eaten by a mouse or rat as has been found in the laboratory, or presumably by an aquatic rodent in the wild state, the mesocercaria, still provided with penetration and cystogenous glands, makes its way to the lungs or muscles of the rodent and then encysts becoming a metacercaria. Further development can only take place if the rodent is itself eaten by a mink, the definitive host, in which the sexually mature flukes are found. In these experiments in the laboratory it was shown by Bosma that three intermediate hosts are obligatory and that if a mink, for instance, eats a frog containing mesocercariae, these develop into metacercariae in the mink and not into mature flukes. In this event it would require the mink to be eaten by another mink to complete the life-cycle: a rodent is accordingly not a paratenic host but an essential or obligatory intermediate host.

Another variation in the life-cycle is one where the cercaria is eaten by the second intermediate host. This occurs in the plagiorchidoid fluke *Triganodistomum mutabile* (Cort) which parasitises the North American cyprinid sucker-fish *Erimyzon sucetta* Lacepède. The cercaria, which lacks a tail and is known as a cercariaeum, develops in sausage-shaped rediae in the snails *Helisoma campanulata* (Say) or *H. trivolvis* (Say) and when liberated sinks to the bottom of the lake where, attached by mucus, it lies until eaten by a planarian or the naiadid oligochaete worm *Chaetogaster limnaei* v. Baer, which is a commensal of fresh-water snails. The cycle is completed when the bottom-feeding sucker-fish eats the worm.

PROGENESIS

Precocious development of the reproductive system has been observed in a number of flukes and in some instances viable eggs are produced before the parasite reaches the definitive host. This condition is known as *progenesis*. A typical instance is that of the metacercaria of *Pleurogenoides medians* (Olsson) which produces eggs while encysted within the second intermediate host *Gammarus pulex* (L.), the definitive host being a frog or toad. Within

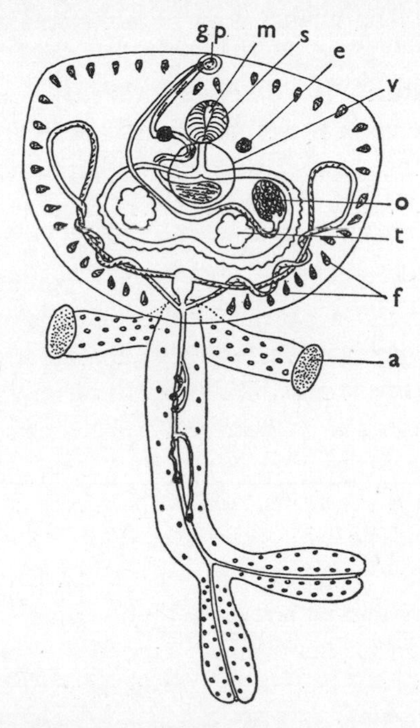

FIGURE 25 *Cercaria patialensis* Soparkar. An example of progenesis where the reproductive system is fully developed in the furcocercaria.
a, adhesive organ; e, eye; f, flame-cells; gp, genital aperture; m, mouth; o, ovary; s, oral sucker; t, testis; v, ventral sucker. (After Soparkar.)

the encysted metacercaria ripe sperms are present and auto-fertilisation takes place. Progenesis has also been described in various cercariae. One of these is the furcocercaria, *Cercaria patialensis* (Fig. 25) first described by Soparkar, which he found within rediae in the fresh-water snail *Melanoides tuberculata* in the Punjab in India, and which has been found in the same species of snail in Ceylon and in *M. anomala* in the Congo. This cercaria has an unusual shape, with a thin, flattened body expanded laterally and with its two intestinal caeca united posteriorly, containing within the region thus encircled the mature reproductive system, while at the base of the tail is a pair of arm-like processes. A slightly larger and more advanced stage of the same parasite in which the forked-tail is lost has been found under the scales of fresh-water fishes in Ceylon, one of these being the green snakehead *Ophiocephalus punctatus* Bloch. This is considered by Crusz and his co-workers to be the progenetic metacercaria of *Transversotrema patialensis* (Soparkar) and they have been successful in completing the life-cycle without the intervention of another host. This stage is thin, flattened and expanded laterally like the body of the cercaria and this shape and the small size of one millimetre broad and half a millimetre long enable it to be accommodated underneath fish-scales.

Although sexual maturity is only reached in the case of *T. patialensis* in a vertebrate host, the cercaria is well advanced towards this in the molluscan host. But there are other records where adult, reproducing flukes have been found in a molluscan host. One of these is *Proctoeces subtenuis* (Linton), described by Freeman and Llewellyn as a fully developed, un-encysted, reproducing digenean in the kidney of the bivalve mollusc known as the mud-hen, *Scrobicularia plana* (Da Costa). Other records of this same fluke are from marine fishes belonging to the families Labridae and Sparidae, indicating that an abridged life-cycle is possible. This confirms the opinion of Serkova and Bychowsky, who found an alternative life-cycle in the fluke *Asymphylodora progenetica* Serkova, a parasite of the roach, *Rutilus rutilus* (L.), which develops first in the gasteropod *Bithynia tentaculata* (L.) and can also reach maturity in another individual of the same species of snail. These then are instances of progenesis but, as Baer suggests, they may be just cases of neoteny, induced perhaps by opportunity and an adequate food supply.

The species discussed in this chapter, as well as representative species of the genera mentioned in the section on classification of Digenea in Chapter 2, are shown in Table 3. This table gives, against each species, the molluscan or first intermediate host, other intermediate hosts when they occur in the life-cycle, and the site of infestation in the definitive host. The variety of definitive hosts in which some of the species are found emphasises the fact that host-specificity is not so marked in digeneans as it is in cestodes and infestation is frequently dependent on ecological habitat and feeding habit.

TABLE 3

DIGENEA—summary of intermediate and definitive hosts of representative species

Species	Molluscan hosts	Other intermediate hosts	Definitive hosts
STRIGEATIDA			
Schistosoma haematobium (Bilharz)	furcocercaria in certain species of *Physa, Bulinus, Planorbis, Physopsis*		capillaries in bladder of man and *Cercopithecus fuliginosus*
Schistosoma bovis (Sonsino)	furcocercaria in certain species of *Bulinus, Physopsis, Limnotragus*		mesenteric and portal veins of zebu, other cattle, goat, horse, etc.
Schistosoma japonicum Katsurada	furcocercaria in certain species of *Katayana* (Japan), *Oncomelania* (China)		mesenteric and portal veins of man, dog, cat, pig, etc.
Schistosoma mansoni Sambon	furcocercaria in certain species of *Planorbis, Biomphalaria*, etc.		mesenteric and portal veins of man, experimentally in laboratory animals
Alaria alaria (Goeze)	furcocercaria in *Planorbis vortex*	mesocercaria in frogs	small intestine of dog, fox, cat, etc.
Alaria mustelae Bosma	furcocercaria in *Planorbula armigera*	mesocercaria in frogs (*Rana clamitans, R. catesbiana*, etc.) metacercaria in mice and rats (experimentally)	intestine of mink and weasel
Strigea vaginata Brandes	furcocercaria in *Spirulina mellea, S. anatina*	mesocercaria or pseudodistomulum in frogs (*Hyla rubra, H. crepitans*) metacercaria or tetracotyle in fish (*Callichthys*, etc.)	hawk-eagle, king-vulture, seriema, etc.

TABLE 3—*continued*

Species	Molluscan hosts	Other intermediate hosts	Definitive hosts
Diplostomum spathaceum (Rudolphi)	cercaria in *Limnaea* spp.	metacercaria in various fresh-water fishes	intestine of gulls (*Larus* spp.)
Transversotrema patialensis (Soparkar)	furcocercaria in *Melanoides tuberculatus* (India, Ceylon), *M. anomala* (Congo)		under scales of green snake-head (*Ophiocephalus punctatus*) in Ceylon progenetic cercaria in *Melanoides tuberculatus*
Cyclocoelum mutabile (Zeder)	cercaria in *Limnaea ovata*		air-sacs of aquatic birds (Anseriformes and Charadriiformes)
Leucochloridium macrostoma (Rud.)	encysted cercariae within sporocyst in various slugs and snails		rectum of various birds (Passeriformes, Columbiformes, Galliformes, Ralliformes, Piciformes)
Proctoeces subtenuis (Linton)	*Scrobicularia plana*		in fishes (Labridae, Sparidae) and in bivalve, mud-hen (*Scrobicularia plana*)
ECHINOSTOMATIDA			
Himasthla quissetensis (Miller et Northrup)	cercaria in *Nassa obsoleta*	metacercaria in *Mya arenaria* and other lamellibranchs	intestine of gulls (*Larus* spp.)
Parorchis acanthus (Nicoll)	cercaria in dog-whelk (*Nucella lapillus*)	metacercaria in mussels (*Mytilus*) and other molluscs	*bursa Fabricii* and rectum of gulls (*Larus* spp.) and crow (*Corvus corone*)
Philophthalmus palpebrarum Looss	not known		conjunctival sac of crow (*Corvus cornix*)

TABLE 3—*continued*

Species	Molluscan hosts	Other intermediate hosts	Definitive hosts
Paramphistomum cervi (Zeder)	cercaria in various snails (*Physa, Bulinus, Planorbis*	encysts on vegetation	rumen of cattle (zebu, buffalo, deer, antelope etc.)
Fasciola hepatica L.	cercaria in *Limnaea truncatula* and other *Limnaea* spp., *Physa acuta*	encysts on vegetation	sheep (type host), other herbivorous mammals, also man
RENICOLIDA			
Renicola brantae McIntosh et Farr	cercaria (rhodometopa) in *Turritella* sp.		kidney of Canada goose (*Branta canadensis*)
PLAGIORCHIDIDA			
Dicrocoelium dendriticum (Rudolphi) = *D. lanceatum* Stiles et Hassall	xiphidiocercaria in *Helicella itala, H. candidula, Zebrina detrita, etc.*	metacercaria in ants (*Formica fusca*, etc.)	bile-ducts and gall-bladder of cattle, sheep, deer, rabbits, dog, cat, man
Haplometra cylindracea (Zeder)	cercaria in *Limnaea ovata, L. stagnalis, Ilybius fuliginosus*	metacercaria in larva of phantom midge (*Chaoborus*)	lungs of frogs (*Rana* spp.) and toads (*Bufo* spp.)
Triganodistomum mutabile (Cort)	cercariaeum in *Helisoma campanulatum, H. trivolvis*	metacercaria in oligochaete (*Chaetogaster limnaei*) or planarian (*Planaria* sp.)	intestine of sucker-fish (*Erimyzon sucetta*)
Asymphylodora progenetica Serkova	*Bithynia tentaculata*		intestine of roach (*Rutilus rutilus*) also in another individual of *Bithynia tentaculata* (its molluscan host)
OPISTHORCHIDIDA			
Heterophyes heterophyes (Siebold)	lophocercous cercaria in *Typanotomus microptera, Melania tuberculata*, etc.	metacercaria in various fishes (*Mugil, Tilapia, Barbus, etc.*)	small intestine and caecum of mammals (man, cat, dog, bat, etc.) and birds (black kite, white pelican)

TABLE 3—*continued*

Species	Molluscan hosts	Other intermediate hosts	Definitive hosts
Cryptocotyle lingua (Creplin)	lophocercous cercaria in periwinkle (*Littorina littorea*)	metacercaria in various fishes (*Cottus*, *Gobius*, *Labrus*)	small intestine of mammals (cat, dog) and birds (gulls, ducks, skua, etc.)
Opisthorchis tenuicollis (Rud.) =*O. felineus* (Rivolta)	lophocercous cercaria in *Bithynia leachi*	metacercaria in various fishes (*Tinca tinca*, *Idus melanotus*, etc.)	bile-ducts of fish-eating mammals (cat, dog, fox, man, glutton, grey seal and porpoise (*Phocaena phocaena*))
Clonorchis sinensis (Cobbold)	cercaria in *Melania hongkongensis*, *Bulinus striatulus japonicus*, etc.	metacercaria under scales and in muscles of cyprinid fishes, also some Percidae, Centrarchidae, etc.	bile-ducts of fish-eating mammals (cat, dog, man)
Gonocerca phycidis (Cobbold)	not known		stomach of various fishes in N. Atlantic and N. Pacific (fork-beard (gadoid), halibut (pleuronectid), mackerel (scombroid), etc.)
Bathycotyle coryphaenae Yamaguti	not known		gills of dolphin (*Coryphaena hippurus*)
Halipegus occidualis Stafford	cystophorus cercaria in *Physa parkeri*, *Ph. sayii*	infestive mesocercaria in *Cyclops vernalis*, etc. metacercaria in nymph of dragonfly (*Libellula incesta*)	buccal cavity of frogs in Canada (*Rana catesbiana*, *R. clamitans*)
UNALLOCATED SPECIES			
Otiotrema torosum Setti	not known		branchial cavity of squaloid fishes (*Squalus*, *Lamna*)
Paronatrema mantae Manter	not known		skin of manta-ray (*Manta birostris*) in tropical waters

8 *The Didymozoonidea*

Cyst-dwellers with a tendency towards unisexuality.
BEN DAWES

Whether the Platyhelminthes considered under this heading are treated as a class or a sub-class, the fact remains that they comprise an isolated group of parasites of fishes whose characters have been epitomised, as above, by Dawes. The two conditions, living in a cyst and showing a tendency towards unisexuality are, in a way, related to each other. Didymozoonids are found in cysts, usually in pairs although more may be present intertwined and crowded together. In some species both partners are similar and hermaphrodite, in others, although still hermaphrodite one partner shows a greater development of the female, and the other of the male reproductive system, while the extreme is met with where one individual functions entirely as a female and the other as a male.

The cyst is produced by the host as a reaction to the presence of the parasites and a site somewhere in the head is usually favoured although some cysts are found in the wall of the gut and others under the skin at the base of the fins. The more usual sites are in the pharynx, on the gill arches or on the gills themselves, in the tissues of the buccal cavity or even in the orbit. With two exceptions, that of a carp in the River Nile, and that of a carp in Japan, the hosts are marine teleostean fishes.

The shape of the animal varies from one species to another. The hermaphrodite *Nematobothrium* van Beneden, for instance, is exceedingly thin and flattened, and when removed from a cyst in the gills of a sun-fish (*Mola mola* (L.)) and unravelled may measure four feet in length. On the other hand the greater part of the body of *Didymozoon* Taschenberg and *Didymocystis* Ariola is swollen, only a short anterior region remaining slender. A variation in the swollen shape is seen in another hermaphrodite species *Didymocylindrus* Ishii, where, as the name suggests, there is a cylindrical body and from the middle of this there arises the narrow fore-body. A similar general arrangement is seen in *Neodiplotrema* Yamaguti, but in this genus part of the hind bodies of the two members of the cyst are fused together, while in *Glomeritrema* Yamaguti the thin bodies become inextricably entangled and actually fuse to form a spherical mass. *Wedlia* Cobbold in common with some other genera is completely gonochoristic, because one of the hermaphrodite individuals develops into a male and the other into a female, there being suppression of the opposite sexual organs in each. There is also a very marked sexual dimorphism, the male being smaller than the female and lodged in a hollow in her side (Fig. 10).

THE REPRODUCTIVE SYSTEM

It is difficult to generalise with regard to this system since there are exceptions to most characters, but in general the reproductive glands are long and slender and not greatly branched, but the system differs from that of the Digenea or Monogenea in the presence of an unpaired vitelline gland. There are usually two testes, though sometimes four, but in *Wedlia* where the sexes are separate there is only one which is compact and rounded (Fig. 10B).

The problem of the mechanism of sex determination in gonochoristic forms is one that awaits solution, but observations on *Didymocystis* suggest that the first parasite to be established develops normally while the development of the partner that joins it later, influenced by the parasite already established, gives rise to a form in which the male system predominates. This is not a unique phenomenon for there are instances in other groups of animals where the development of the sexual organs of one individual is influenced by the presence of an older individual. The echiuroid *Bonellia viridis* is one of these, the slipper-limpet *Crepidula* Lamarck is another, but a more precise parallel may be that of parasitic, cryptoniscid Isopoda where the first individual to be established develops as a female and influences the development of the second which becomes a male.

DIDYMOZOON FACIALE

An account of one hermaphrodite species which embodies most of the common characters should suffice to give a picture of the group, and one of the most readily available species is *D. faciale*, which has been fully described by Baylis (Figs. 9 and 26). The adults are frequently found living in small yellowish cysts measuring from 3 to 7 mm in diameter in the skin of the face or of the head behind the eye of mackerel, *Scomber scombrus* L. The number of individuals in a single cyst varies for there may be two or more present, as many as 16 having been recorded, and they are frequently unequal in size, the largest measuring, when uncoiled, 16 mm long and the smallest less than 1 mm. They differ likewise in shape as they are crowded together within the cyst. The usual form is that of a worm with a thin, slightly flattened, anterior region which terminates in an oral sucker, with the common genital aperture situated on a small papilla, lying close to the sucker. The remainder of the body, swollen and bent on itself, contains the greater part of the extensive and much-convoluted uterus, the diverticula of the gut and the thread-like reproductive glands (Fig. 26).

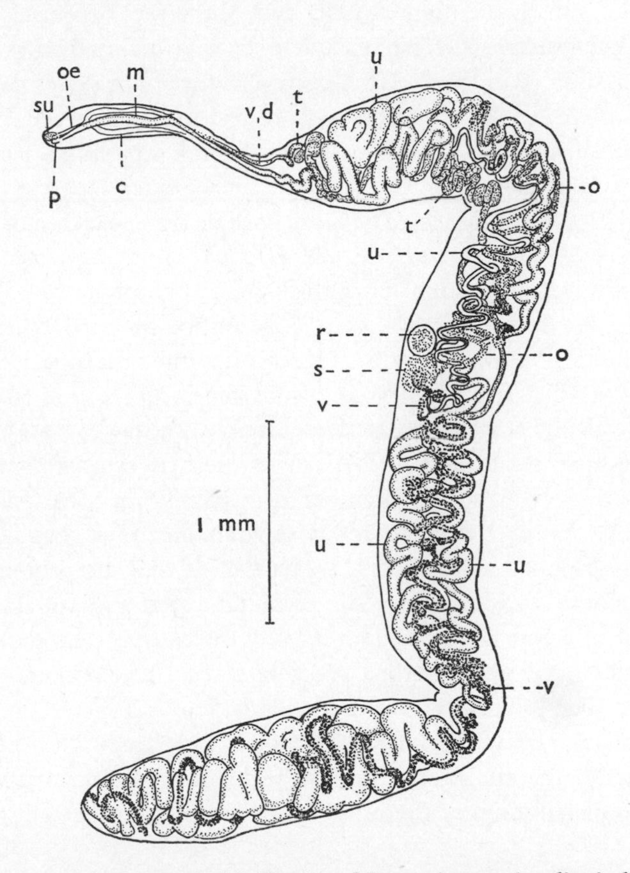

FIGURE 26 *Didymozoon faciale* Baylis. Mature hermaphrodite individual from facial cysts of mackerel, *Scomber scombrus* L.
c, intestinal caecum; m, metraterm; o, ovary; oe, oesophagus; p, genital papilla; r, receptaculum seminis; s, shell-gland; su, sucker; t, testis; u, uterus; v, vitelline gland; vd, vas deferens. (Adapted from Baylis.)

The eggs are small and operculate, measuring only 0·02 mm in length and embryonated eggs are present containing non-ciliated embryos which are armed with a double ring of minute spines at one end (Fig. 9). How much longer they grow before hatching is not known, but in another species the embryo grows in length to become bent on itself within the egg-shell before eclosion when a double ring of alternating larger and smaller spines is seen encircling an oral sucker. The actual penetration by the larva has not been observed in *D. faciale*, but in *Nematobothrium sardae* G. A. et W. G. MacCallum, which is a parasite of the short-finned tunny or bonito (*Sarda sarda* (Bloch)), another member of the mackerel family, it is claimed by Grabda that the larva, after gaining entry into the body, settles down in a blood-vessel. This is the nidus for the cyst, the vessel becoming obliterated as the host encapsulates the parasite, which, it is maintained, accounts for the vascularisation of the wall of the cyst, a feature of many cysts.

There are three significant points with regard to these parasites. The first concerns the larva which differs so much from every other platyhelminth larva that one must conclude that the Didymozoonidea separated early from the ancestral Platyhelminthes. The larva with its double crown of alternating small and large hooklets or spines is a larva equal in phylogenetic importance to the miracidium, coracidium, oncomiracidium, lycophore or cotylocidium. Furthermore, the development is direct. Secondly, they are all adapted to an encysted condition within the host. The third point is the dioecious character of some genera which is obviously not of the same order as that of the schistosomes where all the furcocercariae developing from one egg are of the same sex. As one finds stages intermediate in nature between completely hermaphrodite forms in which each member of a pair is identical and the other extreme where the sexes are completely separate one must conclude that sex-determination is governed by external factors. If they are protandrous, that is hermaphrodites in which the male elements ripen and are shed first, the first member of a pair will develop normally, first testes then ovary, while in the co-partner development stops short after the development of testes, a phenomenon seen to occur elsewhere.

Table 4 shows a list of the species discussed, the nature of the reproductive system of each, the site of the cysts, the kinds of hosts infested and the geographical distribution of these.

TABLE 4

DIDYMOZOONIDEA—summary of representative species

Species	Site of cyst	Hosts	Geographical distribution
Nematobothrium molae Maclaren	in gills, encysted in pairs, hermaphrodite	sun-fish (*Mola mola*)	Eastern N. Atlantic, Mediterranean
Didymozoon scombri Odhner	in gills, 2 or more encysted together, hermaphrodite, tending to separation of sexes	Atlantic and Pacific mackerel (*Scomber scombrus, S. japonicus*) Spanish mackerel (*S. colias*)	N. Atlantic, N. Pacific Mediterranean
Didymozoon faciale Baylis	in skin of face or head, two or more encysted together, hermaphrodite	mackerel (*Scomber scombrus*)	Eastern N. Atlantic (English Channel)
Didymocystis thynni (Taschenberg)	in gills and operculum in pairs, hermaphrodite	tunny (*Thunnus thynnus*)	Mediterranean
Didymocylindrus filiformis Ishii	in gills, encysted in pairs, hermaphrodite	oceanic bonito (*Euthynnus pelamis*), (*Thunnus orientalis*)	N. Pacific (Japan)
Neodiplotrema pelamydis Yamaguti	in gills, encysted in pairs, hermaphrodite	oceanic bonito (*Euthynnus pelamis*)	N. Pacific (Japan)
Wedlia orientalis (Yamaguti) =*Koellikeria orientalis* Yamaguti	in wall of gut, encysted in pairs, distinct sexual dimorphism	oceanic bonito (*Euthynnus pelamis*), tunny (*Thunnus thynnus*), yellow-fin tuna (*Neothunnus macropterus*)	N. Pacific (Japan)
Glomeritrema subcuticola Yamaguti	in pairs in subcutaneous tissue, cyst supplied with capillary network from host, hermaphrodite	spear-fish (*Tetrapterus mitsukurii*)	N. Pacific (Japan)

9 *The Cestodaria*

Le parasite n'est pas un être anormal,
exceptionel, c'est un spécialiste!
JEAN G. BAER, *Le Parasitisme*

It has already been noted (Chapter 2) that many of the animals originally grouped as cestodarians have been weeded out as neotenic cestodes, but there still remain the Amphilinoidea and Gyrocotyloidea. While sharing certain characters such as being unsegmented, lacking a gut, possessing a single set of hermaphrodite reproductive organs and developing from a ten-hooked larva these two sub-classes are otherwise very dissimilar. The Gyrocotyloidea possess a funnel-like adhesive organ while the Amphilinoidea do not; the former have an anterior sucker and the latter have a muscular proboscis, while the ten hooklets of the larva, according to Bychowsky, are not homologous, there being ten similar hooklets in the former, and 6+4 in the latter, the six being homologous, and the four developing as sclerites. It is thus apparent that a proper understanding of cestodarians can only be obtained taking each group separately (Fig. 11A, B).

AMPHILINOIDEA

The amphilinoids are mostly large since fully mature individuals range in size, according to the species, from two to ten inches long. The body, rather like that of a large fluke, is fleshy but flattened, and the surface, devoid of spines, is somewhat rugose in appearance. At one end, which can be designated the anterior end, there is a well-developed muscular proboscis richly supplied with proboscis glands, while at the other there are the openings, close together, of the male and female reproductive systems and the excretory system. The uterus leads to the exterior close to the proboscis.

Reproductive system

The general arrangement of the reproductive system is reminiscent of that of a pseudophyllidean cestode in so far as there is a field of vitelline glands on either side of the body and a posterior ovary. The common vitelline duct joins the oviduct to continue as the ootype surrounded by Mehlis's glands whence it continues as the uterus whose convolutions lead through three limbs, forwards, backwards and forwards again to open in front. The fertilisation canal, or vagina, with its opening at the posterior end of the body, enlarges to form a receptaculum seminis before opening into the oviduct. The male reproductive system, again reminiscent of that of the Pseudophyllidea, comprises a large number of testes distributed in the parenchyma internal to the vitelline glands. The efferent ducts lead into two vasa deferentia which unite to form a common vas deferens into which open numerous prostate glands before opening to the exterior. A cirrus-sac is lacking, but in some species a penis-like papilla marks the external opening of the vas deferens.

Amphilina foliacea

Very little is known of the life-history of amphilinoids. *Amphilina foliacea* (Rud.) (Fig. 11), of which perhaps most is known, is found in the coelom of the sturgeon (*Acipenser*), the site within the host favoured by all except one species. Experiments have shown that the eggs do not hatch in water but only after being swallowed by an amphipod crustacean such as *Gammarus pulex* (L.) and *Carinogammarus roeseli* Gervais. The larva that emerges, known as a *lycophore*, is ciliated, and the ciliated coat is an embryophore, like that of a coracidium, but unlike a coracidium it possesses massive penetration glands. It loses its cilia in penetrating the wall of the gut of the amphipod and it comes to lie in the haemocoel, which is also a site favoured by parasites of many different groups. The penetration glands degenerate, the rudiments of the reproductive system appear, and the larva grows to a length of from 2 to 4 mm when it is a mature and viable larva but it does not lose its embryonic hooks.

The time taken to reach this stage is dependent on the ambient temperature, being about four weeks at a temperature of 20° C but longer when the water is colder. When the infested amphipod is swallowed by *Acipenser* it is digested, liberating the contained larva which bores through the wall of the gut to reach the coelom where it becomes mature. The eggs of *Amphilina* leave the host by way of the abdominal pores.

Other Amphilinoids

Little is known, apart from descriptions of their structure, of other members of this group. Four species of *Amphilina* are known and six other genera have been described which are monotypic. They are found in the body-cavity of sturgeons, of marine bony fishes, in a perch-like, marine fish known as the black sweet-lip, *Pseudopristipoma nigra* (Cuv.) of Ceylonese waters, while in Eastern Australia a species has been found in the lungs of a fresh-water tortoise *Chelodina longicollis* Gray. At first sight it would seem that this is a neotenic cestode which, like *Diphyllobothrium*, has a procercoid stage in the haemocoel of a crustacean but whose life-cycle stops short at the plerocercoid stage in the body-cavity of a fish, a view strengthened by the presence of a ciliated embryophore around the larva which, like that of *Diphyllobothrium*, is lost. (The development of the embryophore and its significance in relation to the absence of a gut is dealt with in the next chapter in describing the embryology of this cestode.) The resemblance stops here, for the lycophore, unlike the coracidium, has large penetration glands reminiscent of those of some miracidia and its ten hooklets are quite diagnostic in spite of Bychowsky's view that there are but six primary hooklets. Although the general arrangement of the reproductive system is like that of a cestode it has features common to those of many other groups of Platyhelminthes, while the arrangement of the muscular layers is quite foreign to that of cestodes.

GYROCOTYLOIDEA

Like Amphylinoidea, the Gyrocotyloidea are hermaphrodite Platyhelminthes with a ciliated larva, the lycophore, characterised by ten pairs of hooklets and large penetration glands. The reproductive organs are not duplicated and there is not a trace of segmentation neither in the excretory system nor in the nervous system. A characteristic feature separating these two groups is the presence at one end of the body of a funnel, wide and short with a much plicate border in *Gyrocotyle* (Plate 2, Fig. 28), but long and narrow and lacking this border in *Gyrocotyloides*. This extraordinary structure, a fleshy funnel, the wide end projecting posteriorly with the aperture of the narrow end in front and lying dorsally, has no counterpart elsewhere unless it is that of the specialised bothria of *Duthiersia* (Fig. 12G) or a tetraphyllidean scolex. As in the scolices of these, it functions as an organ of attachment, in this case to the villi of the spiral valve, a function also of the anterior sucker. Another remarkable feature is the presence of powerful spines with a laminated, concentric structure (Fig. 27A) which are arranged round the anterior sucker and are also found posteriorly. They are embedded in the cortex and are operated by muscles some of which have their origin in the muscular layer separating cortex and medulla. Such bodies are unique in the Platyhelminthes.

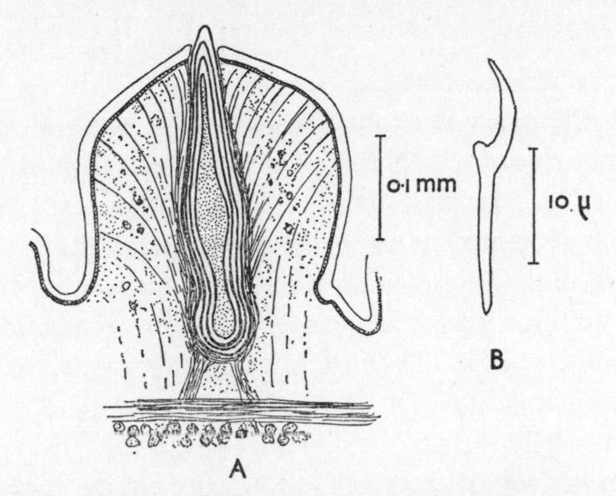

FIGURE 27 *Gyrocotyle urna* Grube et Wagener.
A, spine from crenate margin showing concentric laminae and muscle-attachments. B, hooklet from a lycophore lacking embryophore and found embedded in parenchyma of adult.

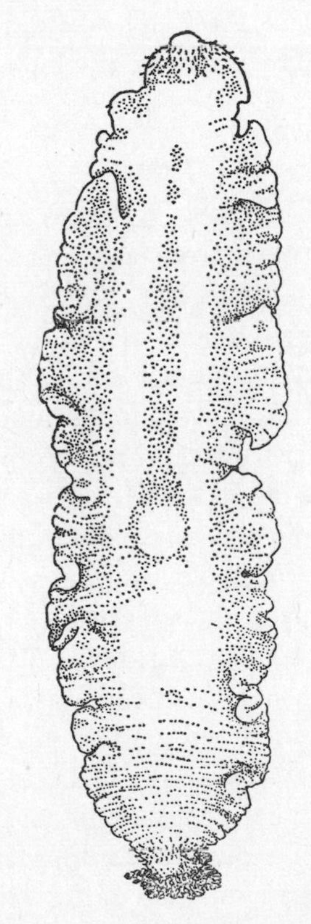

FIGURE 28 *Gyrocotyle urna* Grube et Wagener.

Reproductive system

In a mature individual the uterus occupies the greater part of the medulla, extending from the ovary which is situated in the posterior third of the body, and following a course of tightly folded convolutions to the uterine pore which lies anteriorly on the ventral surface. The vagina leads from its aperture, which is either a separate one or which lies in the end of the uterus, to continue posteriorly, enlarging into a receptaculum seminis prior to joining the oviduct. The vitelline glands are exceedingly numerous, extending into the lateral margins of the body. The testes lie anteriorly in two lateral fields which merge in the mid-line and the vas deferens opens to the exterior dorsally through a penis-papilla, reminiscent of that of the turbellarian *Dugesia*, which lies not far from the vaginal aperture. Thus the genital pores open anteriorly in the gyrocotyloids and at the opposite end of the body in the amphilinoids, which has given rise to much speculation as regards the orientation of these worms. This problem is made more difficult when one considers the nervous system, which is specialised in two ganglia, in the region of the sucker and also at the other end of the body where, in addition, there is a ring-nerve at the base of the funnel and from which nerves arise supplying this body.

Life-cycle

Little is known of the life of these parasites apart from the fact that the lycophore hatches spontaneously from eggs that are voided by chimaeras, and is free-swimming. It has a uniformly ciliated coat, the embryophore; ten similar hooks (Fig. 27B) lie at one end and at the other there open two ducts from the glands that occupy the greater part of the body. It is the presence of the ten hooks that gives an answer to the question of the orientation of the body, because they persist through the larval stage and are still found in the adult in which they lie on a small spherical plug close to the funnel. In the larval monogenean, the oncomiracidium, the hooks lie posteriorly, and in the cestode larva, the procercoid, they are at the end opposite to that of the bothria where they lie on the spherical cercomer; accordingly, if there is any justification in this comparison, one cannot regard the funnel as a modified scolex, and one must consider the funnel to be posterior in position. But this is an academic question and it serves no useful purpose in pursuing it further. Of much greater interest is the fact that cestodarians, like cestodes, lack a gut and that they both derive nourishment from their host through the tegument of the body which is produced into microtriches.

Nothing is known of the dissemination of these parasites, but an experiment by Manter in which newly emerged lycophores were introduced into seawater containing a number of different molluscs, including gastropods and bivalves in addition to fragments of the wall of the gut of *Chimaera*. The lycophores were attracted to the intestinal mucosa into which they buried themselves, without however losing their cilia, and were not attracted to the molluscs. This is a pointer to the fact that if there is an intermediate stage in the life-cycle it is not a mollusc. In scrapings of the wall of the intestine of *Chimaera* Manter has found larvae which have lost their cilia and are barely a quarter of a millimetre long, a fact that seems to indicate the absence of an intermediate host. Larvae are occasionally found, lacking their embryophore, embedded in the tissues of the adult.

RELATIONSHIPS AND ANCESTRY

There are two outstanding characters that link the cestodarians and the cestodes. The first is the absence of a gut at every stage in the life-cycle and the concomitant features, viz. a ciliated embryphore which is lost, the presence of microtriches (Plate 3) in the tegument and the presence of hooklets on the larva, although not restricted to six. The outstanding turbellarian features are the presence of an anterior muscular proboscis in amphilinoids and an anterior sucker in gyrocotyloids. This suggests that both have evolved from a turbellarian stock in the remote past. How remote this is can be gauged from the hosts of gyrocotyloids—chimaeras and selachians. The chimaeras (Holocephali) an ancient and isolated group of fishes and the sharks (Selachii) share common ancestry in pre-Devonian times, over 400 million years ago, and who knows but that gyrocotyloids were a flourishing group of parasites infesting Bradyodonti, Pleuropterygii and even Pleuracanthodii in Palaeozoic times, surviving today in chimaeras and a few sharks. The amphilinoids seem to have diverged early from the ancestral turbellarian stock that gave rise to the gyrocotyloids. In modern times their hosts are sturgeons (Chondrostei), the most primitive known Actinopterygii, but they are also found in teleosts such as cat-fishes (Siluriformes), the sweet-lip (Perciformes), in all of which they are found in the body-cavity, and in a fresh-water tortoise where they infest the lungs. The latter is most likely to have arisen from an ecological association and an invasion of the lungs from the coelom. One has to go far back in time to find a common ancestor of Teleostei and Chondrostei, probably to some pleuropterygian of Devonian times. Of course one cannot say when a particular platyhelminth became a parasite, or when infestation of a particular host took place, but it is significant that the commonest hosts of amphilinoids are sturgeons and those of gyrocotyloids are chimaeras, both groups of hosts isolated and primitive, and it can be argued that cestodarians had origin, as parasites, in the ancestors of these in Palaeozoic times.

A summary of the hosts of representative species is given in Table 5.

TABLE 5

CESTODARIA—summary of hosts of representative species

Species	Site and hosts	Distribution
GYROCOTYLOIDEA		
Gyrocotyle urna (Grube et Wagener)	intestine, chimaera (*Chimaera monstrosa*)	N. Atlantic
AMPHILINOIDEA		
Amphilina foliacea (Rudolphi)	(1) intermediate host haemocoel of crustaceans (*Gammarus pulex*, *Carinogammarus roeseli*, etc.) (2) definitive host body-cavity, sturgeon (Acipenseridae)	Europe
Austramphilina elongata Johnston	body-cavity, lungs, tortoise (*Chelodina longicollis*)	Australia (N.S. Wales)
Gigantolina magna Southwell	body-cavity, black sweet-lip (*Pseudopristipoma nigrum*)	Indian Ocean

10 *The Cestoda or Tapeworms*

So the lone Taenia, as he grows, prolongs
his flatten'd form with young adherent throngs.
ERASMUS DARWIN, *The Temple of Nature*

Tapeworms are parasites in the gut of vertebrates or in derivatives of the gut, as in the bile-ducts of the liver or the intestinal caeca of birds. The body shows three more or less clearly defined regions: first, the scolex which is the organ of attachment, followed by a neck, the region which gives rise to the segments or proglottides which comprise the third region called the strobila. In the strobila one can distinguish the newly formed, or young, proglottides, the most recently formed lying next to the neck, and these show a gradual and increasing complexity until in the mature region of the strobila the reproductive organs are fully formed while the segments in which eggs are found comprise the gravid region. The oldest proglottides, those lying farthest from the scolex, are generally shed, becoming detached one by one, or occasionally a number at a time.

TAENIA PISIFORMIS

A fairly common and representative member of the group is *Taenia pisiformis* (Bloch) (Fig. 29), whose definitive host is a carnivore such as a dog. It has also been recorded from wolves, foxes, jackals and the Cape hunting-dog as well as from lions, tigers, leopards and occasionally the domestic cat. The intermediate host is the rabbit, hare or guinea-pig and even a rat or mouse. The distribution of the worm is world-wide.

T. pisiformis is usually about three feet long although it may attain a length of six feet. The scolex is small and rounded and nearly a millimetre and a half in diameter and it bears a muscular rostellum with some three or four dozen hooks and four suckers. The hooks, arranged as a crown around the

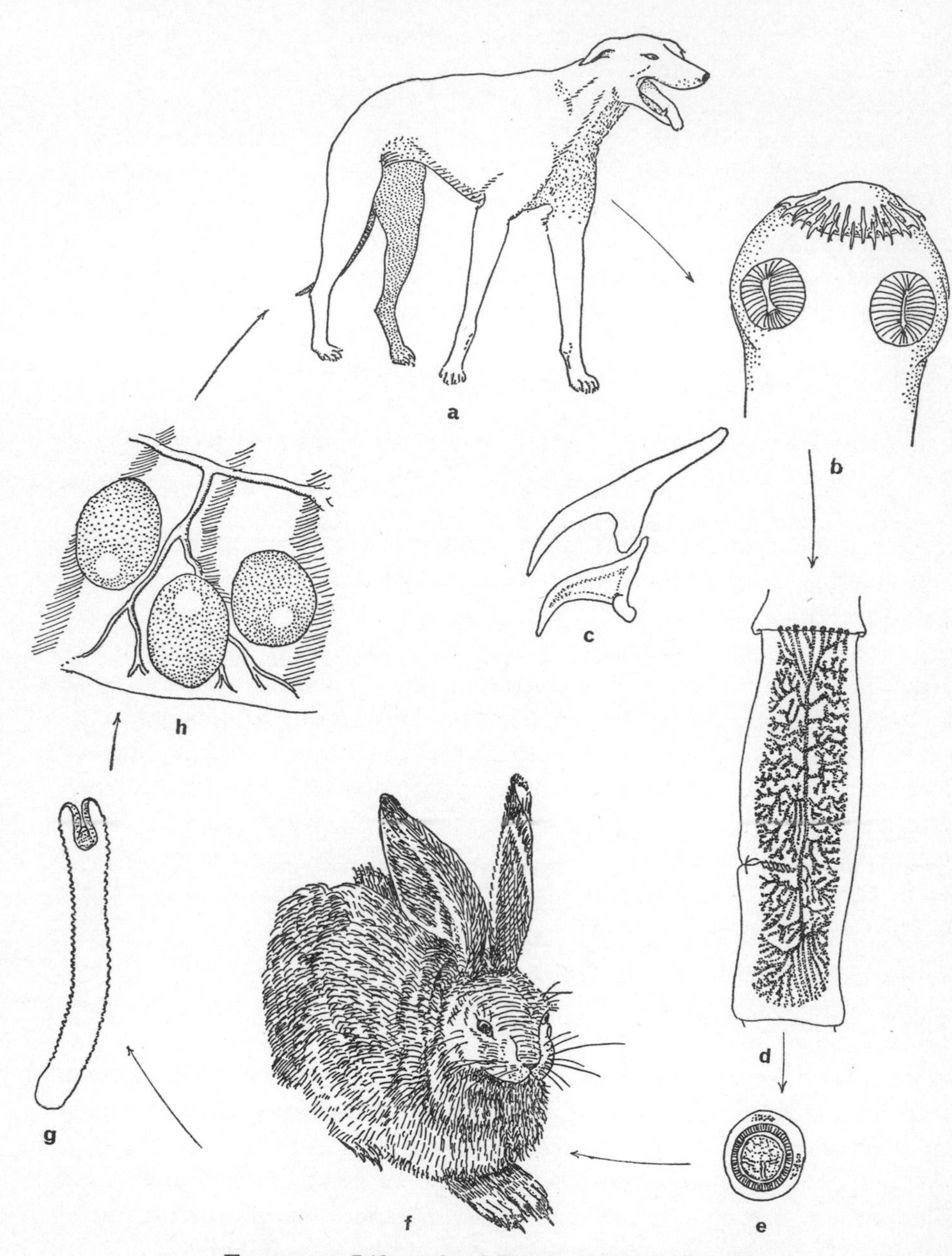

FIGURE 29 Life-cycle of *Taenia pisiformis* (Bloch).
a, dog, definitive host; b, scolex; c, rostellar hooks; d, gravid proglottis; e, egg; f, hare (after Dürer), *Lepus timidus* L.; g, larva from the liver; h, encysted pisiform larvae attached to mesentery.

edge of the rostellum, are of two sizes, the larger hooks with a long cylindrical handle and a rather dumpy blade alternating with smaller, more triangular ones in which the guard of the blade is bifid. The scolex is followed by the neck which merges into the segmented strobila. If, for descriptive purposes, one assumes that the scolex is situated at the anterior end of the worm, then it is at the hind end of the neck that the proglottides are formed, each proglottis gradually increasing in size as the worm grows, and as new proglottides are formed there is a progressive movement of the proglottides away from the scolex. At first undifferentiated, the anlagen or primordia of the reproductive organs appear as small masses of cells; these take form as the genital apparatus which eventually becomes complete in each proglottis.

Reproductive system

The proglottis is flattened, and the two flattened sides are considered as dorsal and ventral, the convention being that the ovary lies ventrally in the parenchyma. On one of the lateral margins is the opening of the genital atrium which receives the cirrus, the eversible terminal part of the vas deferens, and the vagina which is the fertilisation canal. The ovary lies ventrally and posteriorly and is bilobed and from the connection between these lobes there arises the oviduct which joins a duct from the vagina, or more correctly from the receptaculum seminis which is just an expanded part of the vagina. Lying immediately behind the ovary in the posterior part of the proglottis is the vitelline gland, and the vitelline duct from it joins the united oviduct and vagina to form the *ootype*. The latter is a short specialised region surrounded by glands known as Mehlis's glands, and into this region there come ovum, yolk-cells and sperm. It is here that fertilisation takes place and the egg-shell is formed, secretions from Mehlis's glands and from the yolk-cells taking part in its formation. As in the assembly line of a factory, as one article is completed it moves forward, and the elements of the next come into place, so here also the completed egg moves on into the uterus which continues from the ootype. In all taeniids the uterus is a median stem with lateral diverticula. As the eggs pass into it the uterus expands and fills with eggs (Plate 5).

The male reproductive system comprises a large number of testes, small spherical bodies which lie dorsally in the medulla, and their efferent ducts lead towards the centre of the proglottis uniting to form the vas deferens which follows a sinuous and convoluted course becoming expanded and functioning as a seminal vesicle. It leads through the cirrus-sac as the ejaculatory duct and terminates as the eversible cirrus. The cirrus-sac has strong muscular walls and within it, opening into the ejaculatory duct, are prostate glands. The male reproductive system develops in the strobila before the female system and a consequence of this protandry is that sperm from a more anterior region of the strobila can fertilise the ova in a more mature region. This is a simple enough procedure when one realises that cestodes do not lie stretched out and flattened against the wall of the gut, but that, provided with a complex and well-developed muscular system, they can stretch out, contract and otherwise move and twist their bodies while the scolex remains fixed in position.

Apart from the reproductive system there are three other systems worthy of note: the nervous system, the excretory system and the muscular system. The nervous system can be seen in each proglottis as lateral nerves, one lying on each side at the junction of medulla and cortex. These nerves arise from a pair of ganglia in the scolex, from which there also arise nerves to the suckers and to the rostellum. The muscular system comprises layers of muscle-fibres. Immediately beneath the tegument there are two layers, an external one of circular fibres or fibres lying transversely in respect of the proglottis, and internal to this are bundles of longitudinal fibres. The central part of the parenchyma, the medulla, is delimited by layers of transversely arranged fibres lying internally, and external to these there are numerous bundles of fibres lying longitudinally. Dorso-ventral fibres extend through the medulla connecting the muscular layers which separate this region from the cortex (Plate 6).

The excretory cells are not easily seen except in a living specimen when, under the microscope, careful focusing can bring into view the flickering cilia which lie within each of these excretory cells or solenocytes. Ducts from the excretory cells lead into the excretory canals. The excretory canals lie in the lateral region of the medulla, a wide, thin-walled, ventral, excretory canal and a narrow, thicker-walled dorsal longitudinal canal lying on each side, the ventral canals being connected by a transverse canal lying posteriorly in each proglottis.

Life-cycle (Fig. 29)

Segmentation of the ovum starts immediately after fertilisation and before the terminal proglottides are shed each egg-shell contains a six-hooked embryo, the hexacanth embryo or oncosphere, which is surrounded by a radially striated embryophore, the significance of which will be dealt with later in discussing the development of *Diphyllobothrium latum* (L.). The gravid proglottides pass out of the host with the faeces and they either disintegrate, setting free the eggs which may be eaten by a rabbit or hare or other browsing animal, or it may be that the whole proglottis is swallowed.

Under the influence of pepsin in the stomach and trypsin in the intestine,

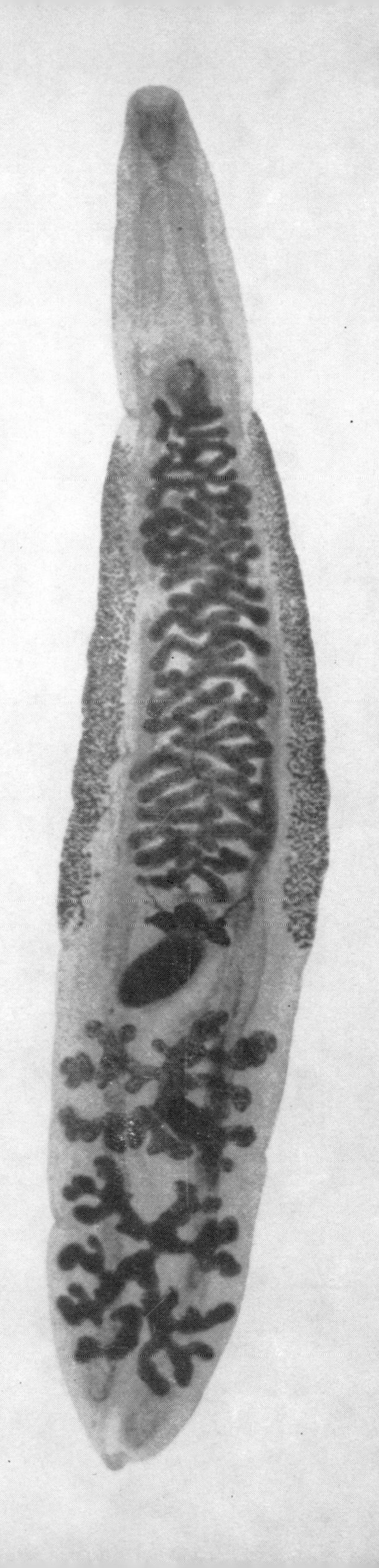

PLATE I. *Clonorchis sinensis* Cobbold, the Chinese liver-fluke, from the bile-ducts within the liver of man and other mammals. Length 14 mm. Photograph by courtesy of Ward's Natural Science Establishment Inc., Rochester, N.Y., U.S.A.

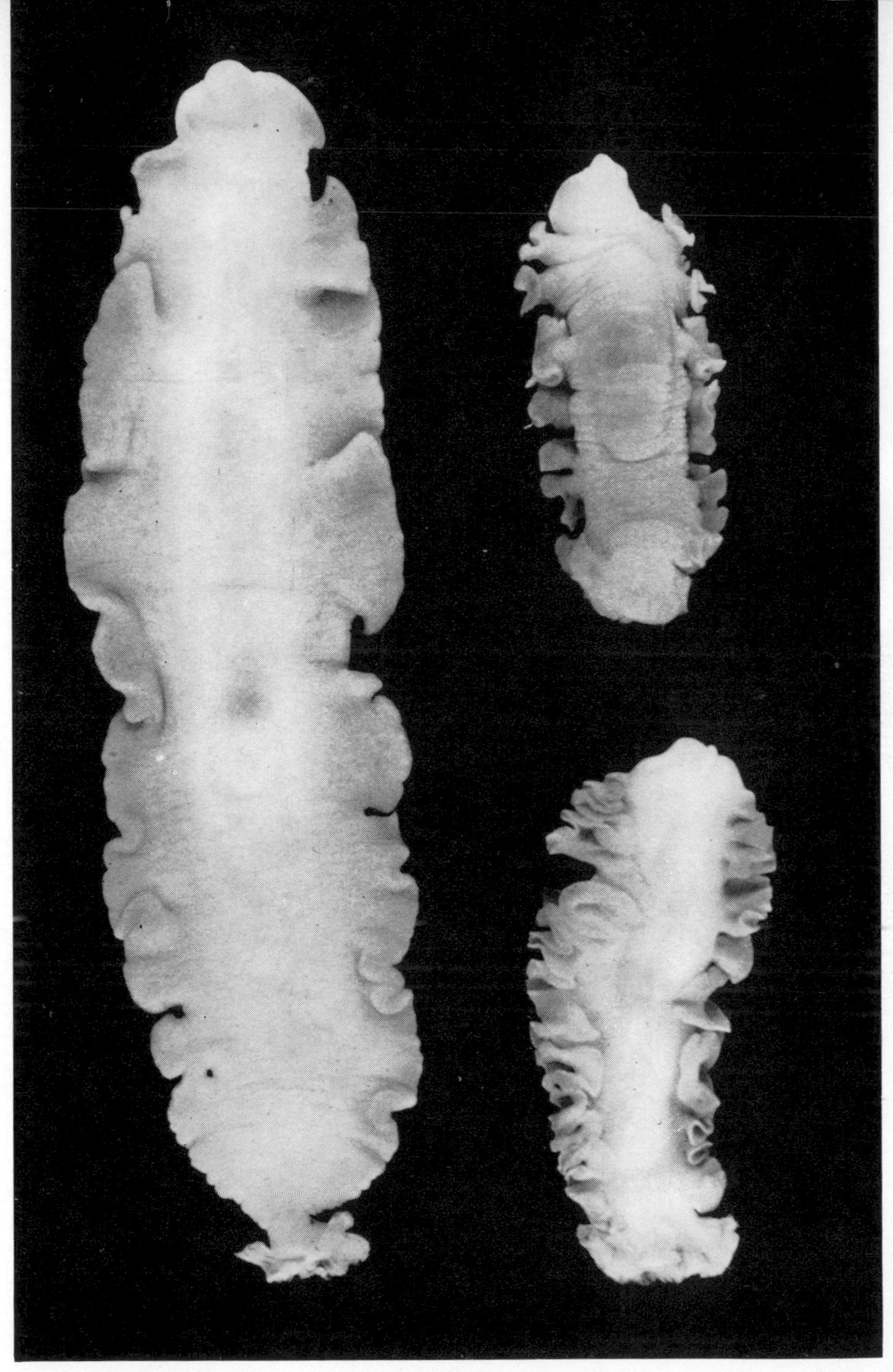

PLATE 2. *Gyrocotyle urna* (Grube et Wagener) from the intestine of *Chimaera monstrosa* L.

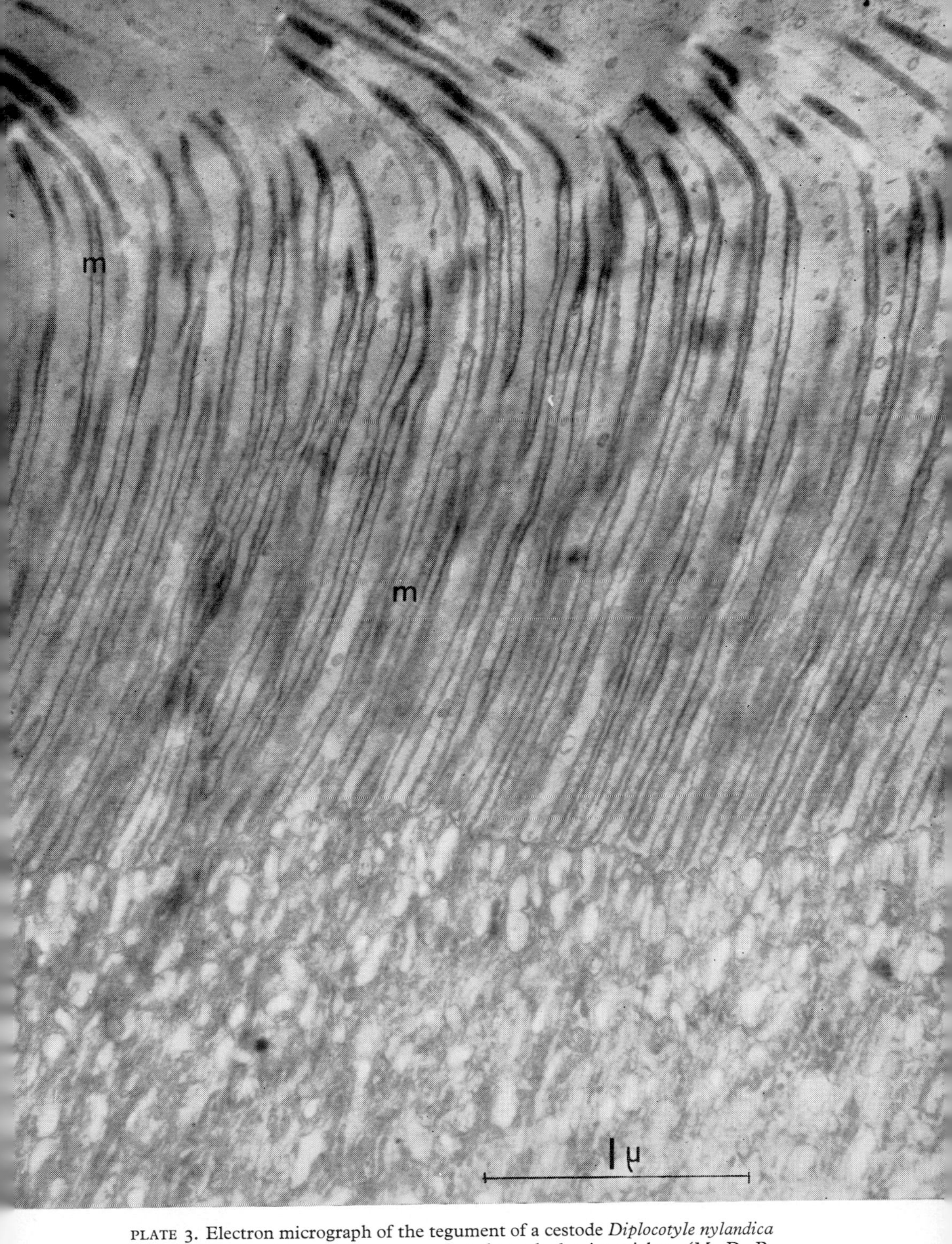

PLATE 3. Electron micrograph of the tegument of a cestode *Diplocotyle nylandica* (Schneider) showing the numerous, densely-packed microtriches. (M. D. B. Burt) × 42,000

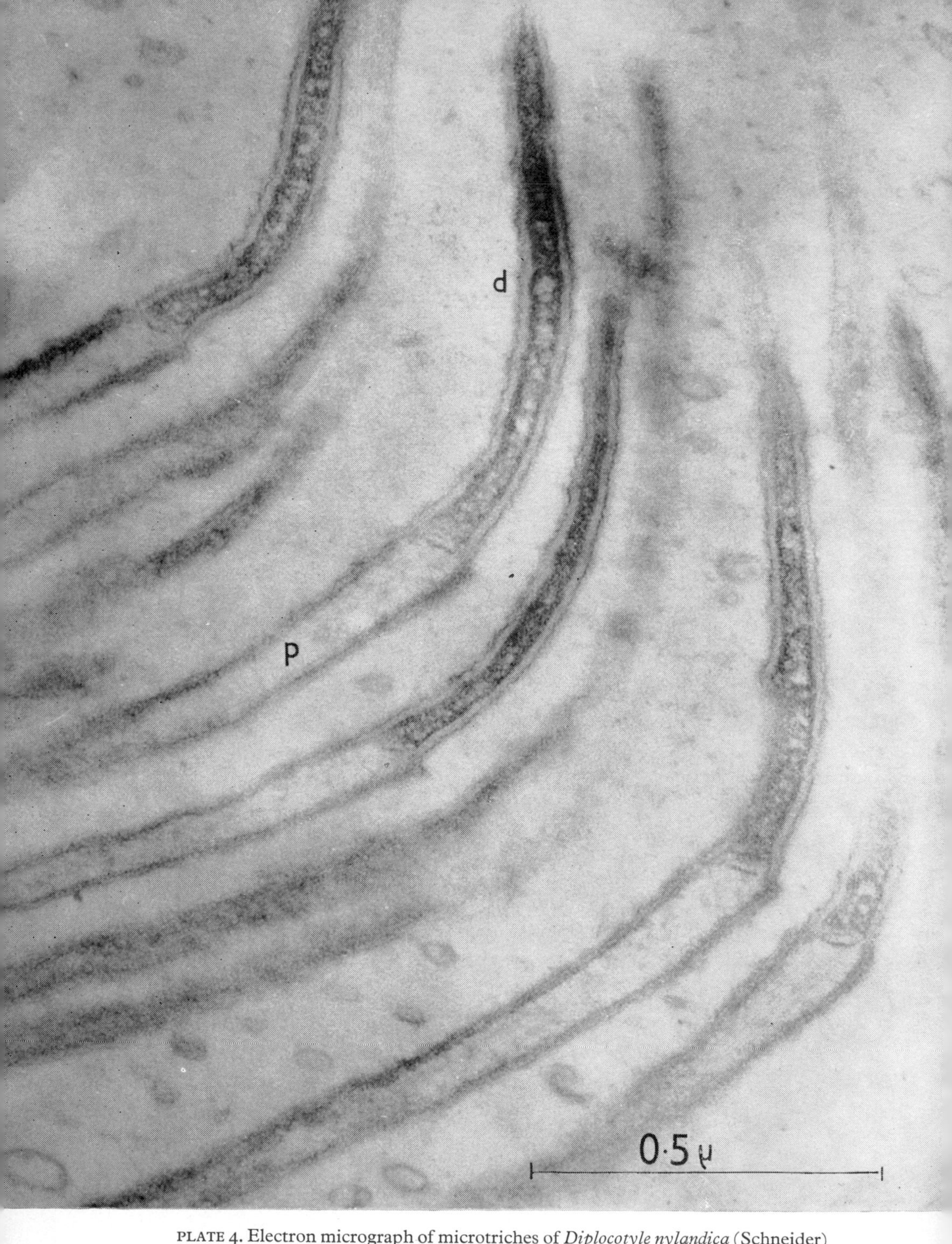

PLATE 4. Electron micrograph of microtriches of *Diplocotyle nylandica* (Schneider) showing the ultrastructure of the tubular proximal (p) and the electron-dense, distal moiety (d). (M. D. B. Burt) × 110,000

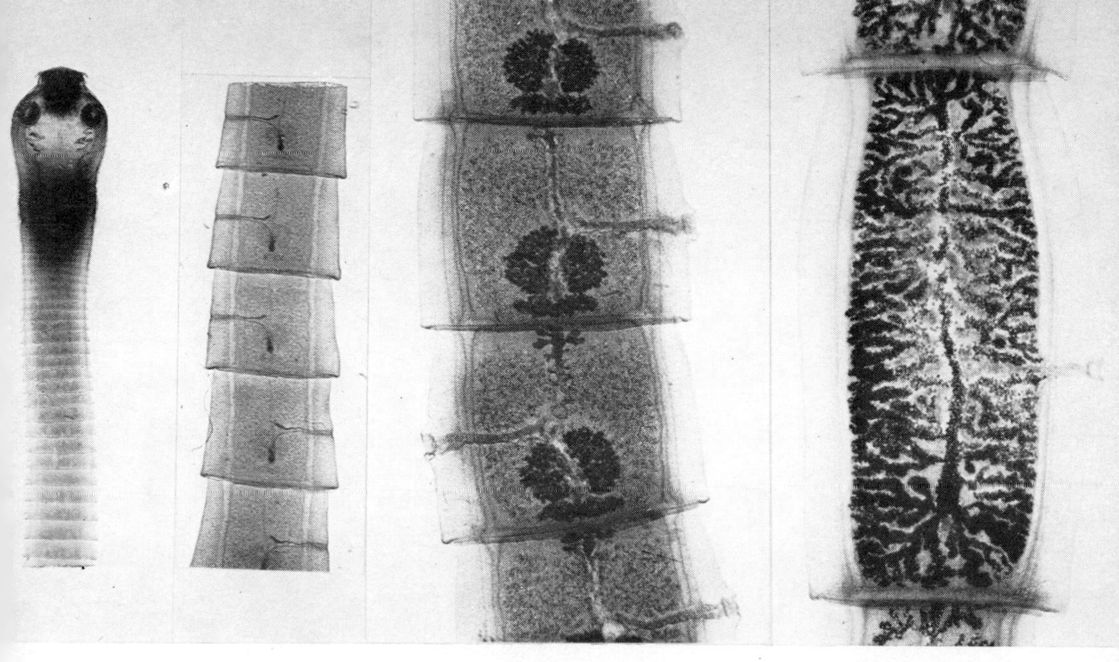

PLATE 5. *Taenia pisiformis* (Bloch), a very common tapeworm of dogs with intermediate stage in rabbits or hares. Scolex, immature, mature and gravid proglottides. (Courtesy Ward's Natural Science Establishment Inc., Rochester, N.Y., U.S.A.)

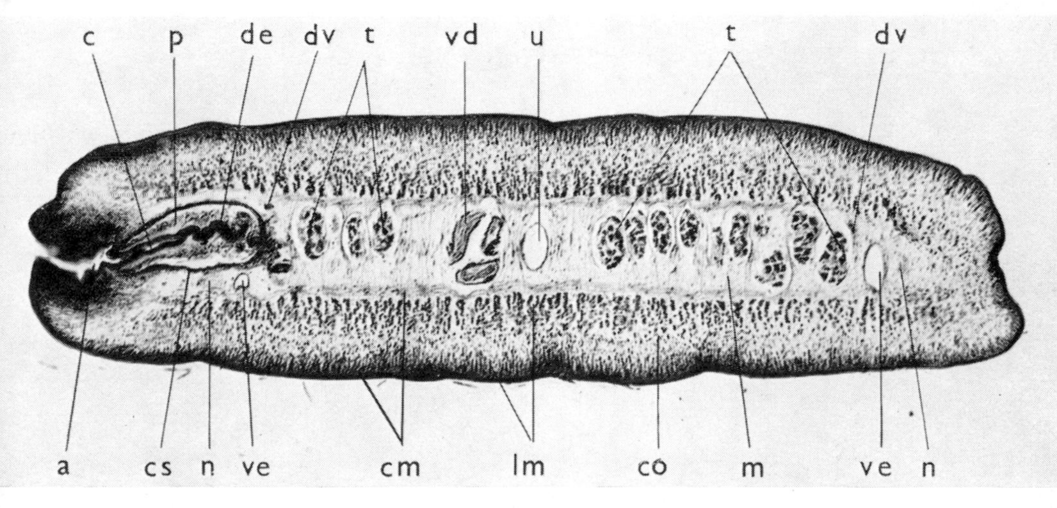

PLATE 6. Transverse section through a mature proglottis of *Taenia pisiformis* (Bloch). a. genital atrium; c. cirrus; cm. circular, or transverse muscle fibres; co. cortex; cs. cirrus-sac; de. ductus ejaculatorius; dv. dorsal, longitudinal, excretory vessel; lm. longitudinal muscle-fibres; m. medulla; n. lateral nerve; p. prostate-gland cells; t. testes; u. uterus; vd. vas deferens; ve. ventral, longitudinal, excretory vessel.

PLATE 7. Coenurus of *Multiceps serialis* (Gervais) from the axilla of a rabbit, cut open to show the scolices which have developed close together in a few rows.

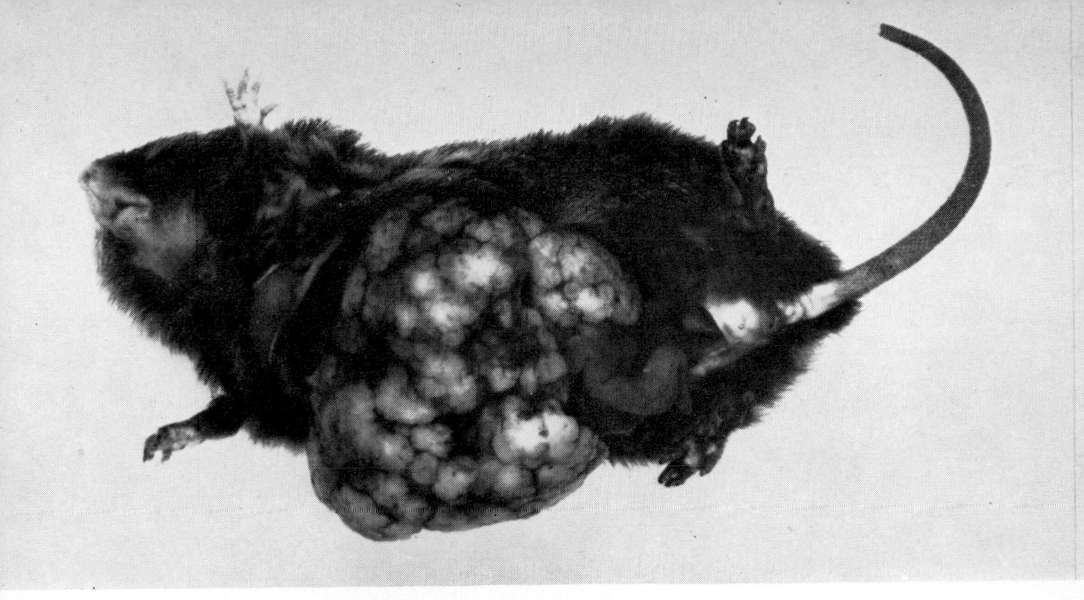

PLATE 8. A Cotton-rat (*Sigmodon*) of North America which has died from echinococciosis. The abdomen is opened to show the heavy infestation of the liver with the multilocular hydatid of *Echinococcus multilocularis*.

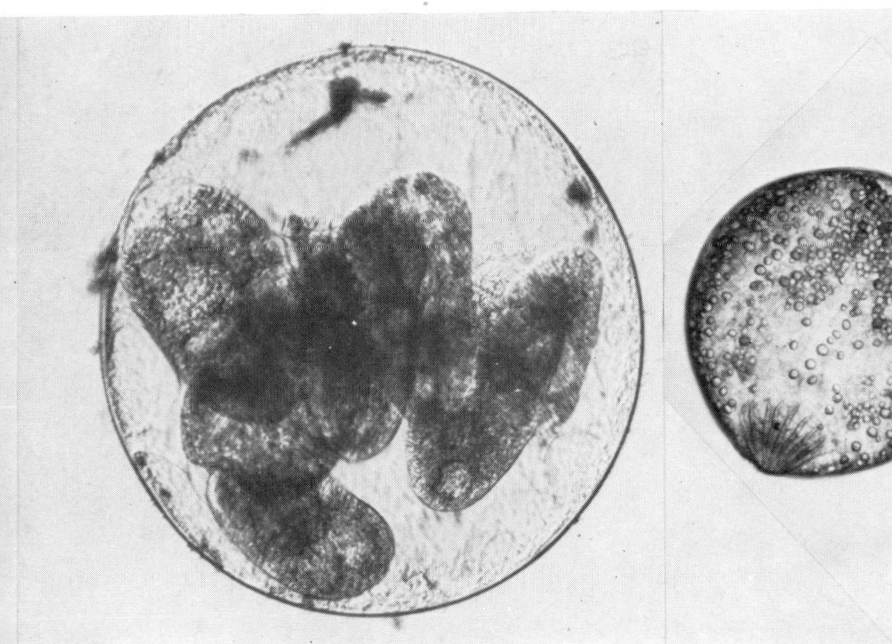

PLATE 9. *Paricterotaenia paradoxa* (Rudolphi). A live polycercus containing ten cysticercoids. (J. S. Scott)

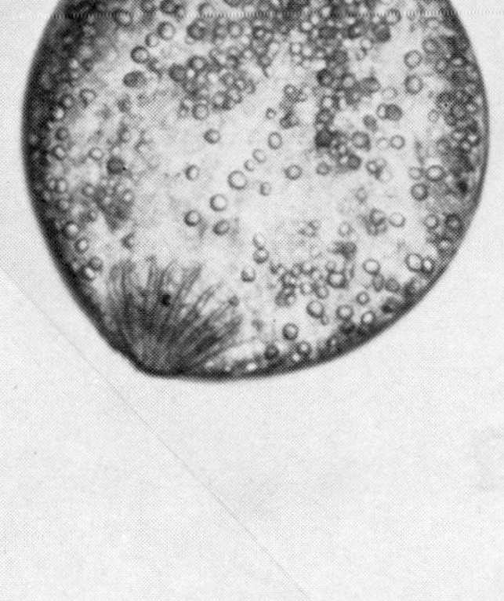

PLATE 10. *Paricterotaenia paradoxa* (Rudolphi). A live cysticercoid removed from a polycercus.

PLATE 11. *Schistosoma japonicum* Katsurada, the causative agent of Asiatic schistosomiasis (*katayama*), whose usual habitat is the venules of the intestinal wall. In the male fluke shown here the gynaecophoric canal, suckers and testes are apparent. (Courtesy Ward's Natural Science Establishment Inc., Rochester, N.Y., U.S.A.)

the outer shell is dissolved and the embryophore breaks up to liberate the hexacanth larva. By means of its hooks the larva penetrates the wall of the intestine and gains a tributary of the portal vein which carries it to the liver. Some may enter the lymphatics and are thus carried to the right heart and the lungs where they perish, others may end in mesenteric ganglia, but it is those that reach the liver that develop normally.

Development is fairly rapid for within a couple of days the site of infestation in the liver is made manifest by the presence of white spots, due to the accumulation of leucocytes, and by the end of a fortnight the oncosphere has lost its hooks and become a small vesicle. This body grows in length to about a centimetre appearing as a narrow strand little over a millimetre in thickness and full of parenchymatous tissue. A constriction appears about the middle of the strand and the moiety which we can designate as posterior separates and eventually disintegrates, while at the anterior end of the other part an invagination takes place at the bottom of which a scolex eventually forms. This larva is very active and within a month from entering the liver it leaves it to wander about in the coelom for several days before attaching itself to the peritoneum or mesentery where it produces an inflammatory condition, eventually coming to lie under the peritoneum or within the mesentery. The parenchyma within the larva degenerates, the cyst so formed becoming full of a clear fluid, and it increases to the size of a pea within a few weeks, and can now be called a bladder-worm or *cysticercus* (Fig. 29h). The invagination noted earlier is directed towards the centre of the bladder, and at the inner end of it a scolex develops as an invaginated body, with the rostellum at the bottom of the diverticulum and four suckers, facing towards its cavity lying close to it. This is the condition of the bladder-worm until it is eaten by the definitive host in whose stomach the external envelope is digested, but the scolex remains invaginated until the larva arrives in the duodenum, when, under the influence of bile and pancreatic juice, evagination of the scolex and neck takes place. The rostellum and suckers are immediately active and the scolex becomes firmly attached to the wall of the gut. One frequently notes that some animals will tolerate some parasites and not others, but in the case of *T. pisiformis* the boot is on the other foot for this tapeworm will only tolerate certain hosts. It has been found that certain salts of glycocholic acid, present in the bile of most animals, are inimical and even toxic for *Cysticercus pisiformis*, and that these salts are absent from the bile of members of the dog family, and this determines that dogs and their kind are definitive hosts.

Within a fortnight of ingestion of *C. pisiformis* the young tapeworm is an inch long, and by the end of two months sexual maturity has been reached in the terminal segments of the strobila.

DIPHYLLOBOTHRIUM LATUM

This tapeworm is commonly referred to as the broad tapeworm of man, and as it is a pseudophyllidean it differs both in morphology and life-history from *T. solium* L., the solitary tapeworm of man. It is, however, not confined to man but is a parasite of various fish-eating mammals among which dogs, cats, bears, seals and porpoises are common hosts although the list of other animals from which it has been recorded is an extensive one (Fig. 30). The record of intermediate hosts is no less formidable, for as a first intermediate host, that of the procercoid, there may be certain species of copepods. Every fresh-water copepod is not a suitable host for some kill and digest the coracidium while others are killed by it. Among suitable hosts are *Cyclops strenuus* Lillj., common in Europe, and *Diaptomus oregonensis* Fisch. in North America. The plerocercoid is found in an even greater number of hosts, for these are fishes belonging to many different families and the records include salmon, trout, pike, perch, carp, white-fish, loach and eels. Adults are found in the small intestine, procercoids in the haemocoel of copepods while the plerocercoids are usually found in the muscles of fishes, but other organs are recorded and these seem to vary according to the particular fish infested.

D. latum may attain a considerable size, for its length is usually 20 to 30 feet, but cases are on record of worms 40 feet long, and in its length there may be over 3,000 segments. The scolex is elongate and pointed, compressed from side to side and with a dorsal and ventral adhesive groove or *bothrium* (Fig. 12D). In the youngest proglottides no trace of reproductive organs can be seen, but they are apparent in the older proglottides and in those in the region of the 600th proglottis both male and female systems are fully developed and can be considered as mature. Mature proglottides are much broader than long with projecting posterior margins. In a mature proglottis there are two apertures in the mid-ventral line, one lying behind the other, the anterior one is the common genital aperture and behind it lies the aperture of the uterus.

Female reproductive system

The bilobed ovary lies ventrally in the medulla and the vitelline glands are disposed throughout the cortex. From the common genital aperture the vagina leads posteriorly to unite with the oviduct and the combined duct receives a duct from the vitelline glands continuing as the ootype which is surrounded by Mehlis's glands. The ootype continues as the uterus which follows a convoluted course and shows four or five conspicuous loops or coils on each side, giving the appearance of a rosette, and thus it leads to the uterine aperture.

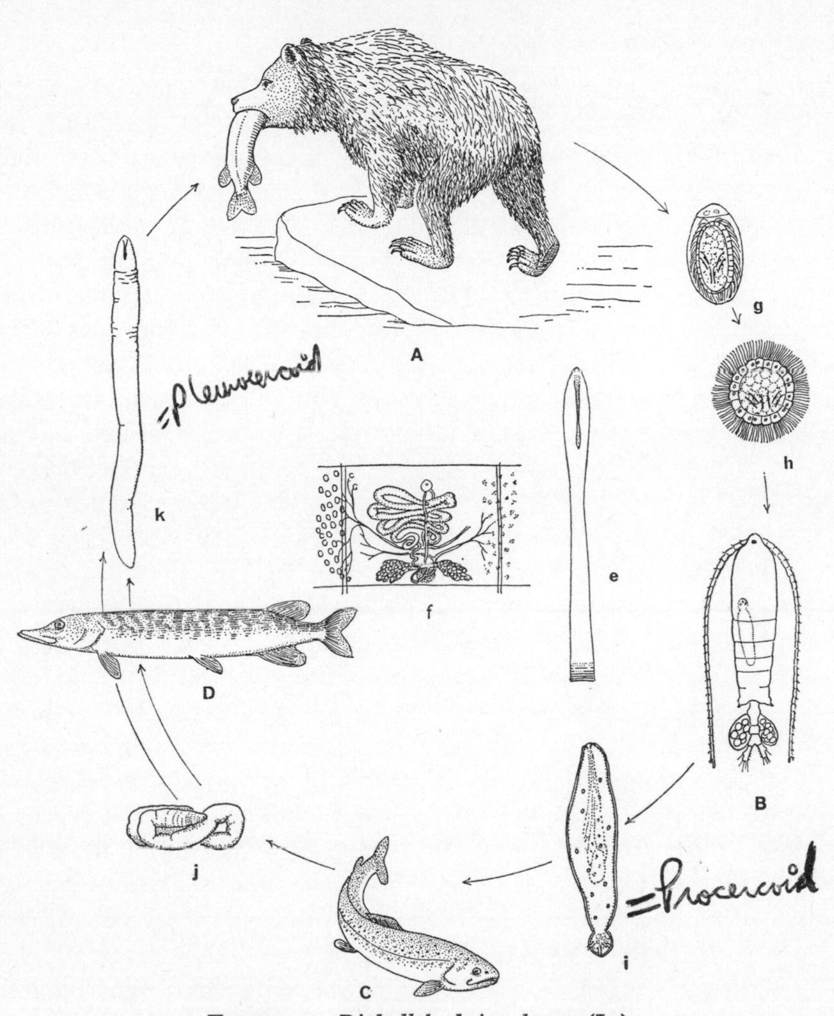

FIGURE 30 *Diphyllobothrium latum* (L.).
The definitive host is a fish-eating carnivore such as a bear, *Ursus* sp. (A) or a man, while the first intermediate host is a crustacean, *Diaptomus oregonensis* Lilljeborg (B) in North America or certain other species of *Diaptomus* or of *Cyclops*. A salmonid (C) may act as the second intermediate host, while the pike, *Esox lucius* L. (D) may participate in the life-cycle as a paratenic host. The scolex (e) shows one of its bothria while the hermaphrodite reproductive organs in the central region of a proglottis are seen in (f). Eggs, each containing a fully developed embryo (g) are passed by the definitive host and from each there hatches the ciliated, free-swimming coracidium (h) which is eaten by the crustacean (B) in which it develops into a procercoid (i). When the crustacean is swallowed by the second intermediate host (C) the procercoid makes its way to the muscles and becomes encapsulated, developing into a plerocercoid (j). The fully developed plerocercoid (k) emerges from the second intermediate host (C) or the paratenic host (D) when eaten by the definitive host and develops into the strobilate cestode.

Male reproductive system

There are from 750 to 800 testes, small spherical bodies which lie dorsally in the medulla in a single field on each side of the proglottis. The efferent ducts unite to form a sinuous vas deferens which runs forwards to continue through the cirrus-sac to open into the genital atrium in front of the vaginal aperture.

Eggs are liberated continuously through the uterine aperture of gravid proglottides and it is only the exhausted proglottides that are eventually shed. The eggs pass out of the host with the faeces and are non-embryonated, further development taking place if they are deposited in fresh water such as a stream or a lake. A free-swimming coracidium (Fig. 30h) eventually emerges, the time of development varying, according to temperature and probably to available oxygen, from about ten days to several weeks.

Development of the coracidium

A knowledge of the embryology of cestodes is necessary if one is to understand the significance of the apparent absence of a gut in this group, a fundamental character that separates the cestodes and cestodarians from all other Platyhelminthes. It is perhaps easier to do this in *D. latum* than in most tapeworms, because the eggs are non-embryonated when laid and one has only to separate some of them in clean, aerated water and examine them as they develop.

Before development starts one can see within the egg-shell the fertilised ovum surrounded by a number of vitelline cells (Fig. 31A) which have now lost many of the granules that characterised them earlier, these having been used during the process of egg-shell formation. The ovum divides, and one of the cells so produced separates from the other, passes between the vitelline cells to lie just within the egg-shell, where it later divides and forms a thin vitelline membrane which surrounds vitelline cells and segmenting ovum. The remaining cell of the ovum divides twice to form four cells or blastomeres (Fig. 31B). One can trace the fate of these blastomeres in subsequent divisions for one of them gives rise to a layer of cells surrounding the products of division of the other three. The result is a solid mass of cells, described embryologically as a *morula*, or mulberry mass, in which the outer layer is derived from one blastomere and the inner cell-mass from the other three and this morula is enclosed, together with the remains of the vitelline cells, within the thin vitelline membrane (Fig. 31C). The outer layer is the epiblast, frequently called the ectoderm, whose cells become columnar, separate from the inner cell-mass and develop cilia becoming the embryophore. The latter contains the inner cell-mass which becomes organised as the oncosphere with its six hooks (Fig. 31D). The development of *Taenia* follows a similar course except that the outer layer or epiblast does not become ciliated but becomes transformed into a radially striated embryophore, homologous with the ciliated embryophore of *D. latum*. The significance of this development lies in the separation of the epiblast from the rest of the embryo and its subsequent loss, the hexacanth embryo being formed of hopoblast and mesoblast, and due to epiboly the hypoblast comes to lie on the outside. Consequently the outside of the adult worm is endodermal, so that the cestodes, far from lacking a gut, possess one in the form of the whole of the external surface and a cestode obtains its food by absorption through this surface. This view is confirmed

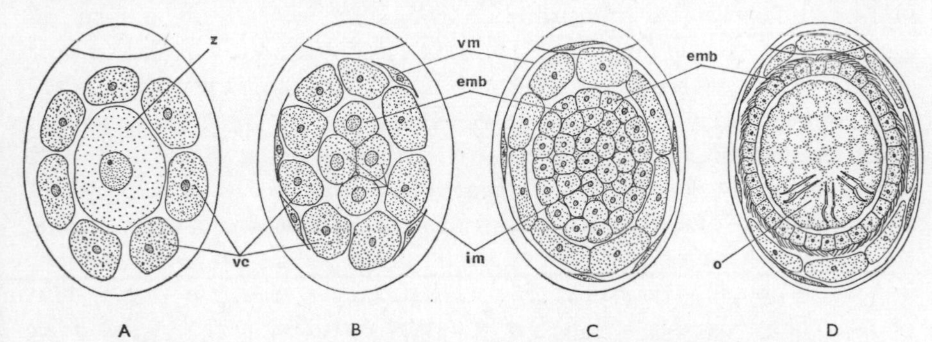

FIGURE 31 Development of *Diphyllobothrium*—semi-diagrammatic. A, Zygote (z) and vitelline cells (vc) are contained within the egg-shell. B, The zygote has divided, one blastomere has given rise to the cells which will form the vitelline membrane (vm), the other has divided twice to form four cells; one of these (emb) will give rise to the embryophore and the other three (im) will form the inner cell-mass. C, Morula stage. The outer layer of the embryo is becoming differentiated as the embryophore (emb) which is a distinct layer surrounding the inner cell-mass. D, Egg just prior to emergence of the coracidium. The embryophore is now ciliated (emb) and the oncosphere (o) has developed from the inner cell-mass and shows six hooks.

by using the electron microscope to study the structure of the tegument and comparing it with a similar study of the fine structure of the surface of intestinal epithelium. This has brought to light the presence of innumerable fine processes projecting from the external surface of tapeworms which have been named microtriches (*μικρός* small, *τριχοι* hairs) (Plates 3, 4). Each microthrix (*θρίξ* hair) is a hair-like body, about 70 mμ in diameter which shows a tubular structure proximally and a slightly narrower, tapering and electron-dense region distally. They are similar in structure to the microvilli now found to be characteristic of internal absorbing surfaces. In addition the discovery that enzymes are secreted through the external surface of tapeworms is another gut-like character. The most recent conclusion on this subject, by François Béguin is, 'Nous pouvons, à titre d'hypothèse, admettre que la cuticle et les celles sous-cuticulaires des Cestodes seraient l'équivalent respectivement de l'ectoplasme et de l'endoplasme des cellules épithéliales intestinales.'

Life-cycle (Fig. 30)

The coracidium is short-lived but if, within about twelve hours from emerging, it is eaten by a small crustacean, such as certain species of *Cyclops* or *Diaptomus*, the ciliated embryophore is shed and the liberated oncosphere bores through the gut-wall to lie in the haemocoel (Fig. 30). Here it develops into a *procercoid* in about three weeks and by this time it is about half a millimetre long with a protrusible appendage at one end and a spherical appendage containing the six hooks at the other. When the copepod infested with the procercoid is eaten by a fish it is digested and the spherical appendage of the procercoid with its six hooks becomes detached, while the remainder bores through the wall of the gut and starts wandering in the body. It eventually settles down in some organ, such as the liver or muscles, which becomes the site of further development and in about six weeks it has become a fully formed plerocercoid which is no longer a larva but a very juvenile cestode of about a centimetre long whose scolex possesses bothria. This plerocercoid is capable of infesting the definitive host, man or some other fish-eating mammal, but if the host of the plerocercoid is eaten by another fish, as for instance if a trout is eaten by a pike, the pike becomes a paratenic host (p. 44) and re-encapsulation occurs, and the plerocercoid penetrates to the same organs as those in which it developed in the first fish. In this way the plerocercoid may pass through a number of hosts, and an old predatory fish such as a pike may have thousands of plerocercoids encapsulated in its muscles.

The distribution of this parasite naturally depends on the distribution of certain copepods and on the availability of fish as a food, but as far as man is concerned a further factor must be considered as to whether he prefers his fish raw or well cooked. The condition of possessing *D. latum* as a parasite is known as diphyllobothriasis or by its synonym bothriocephalosis, and it is more common in some regions than in others. In Europe the most important regions are north-west Russia and the countries bordering on the Baltic, including Finland and extending as far as Denmark and Holland. There are also instances in Bavaria, Hungary, the lower reaches of the Elbe and the delta of the Danube. Other regions are determined by the presence of lakes, such as the Swiss lakes, Lake Tiberias, Lake Bourget in France and the Great Lakes in North America. The dog as a host merits special attention on account of his close association with man, for in East Prussia as many as 50 per cent of dogs are infested compared with 30 per cent in

Canada. But in Canada there exists a natural infestation in bears, and it is possible that some of the human infestations may have had their origin in this source by way of lumbermen.

PARICTEROTAENIA PARADOXA

A third example of a cestode is *Paricterotaenia paradoxa* (Rudolphi), an avian parasite of various wading birds (*Charadriiformes*) such as the woodcock (*Scolopax rusticola* L.), oyster-catcher, plovers, sandpipers and snipe. The intermediate host is an earthworm *Allolobophora terrestris* (Savigny), and the definitive host bears to the intermediate host the relationship that obtains in most cestodes, that of predator to prey. It is to be noted that although many different definitive hosts harbour this tapeworm only one species of earthworm is the intermediate host, and that earthworms of other species are not infested, even in regions where the incidence of infestation is high as in woodcock-brakes.

Life-cycle (Fig. 32)

P. paradoxa is usually found in tens of thousands, and sometimes even hundreds of thousands, in the intestine of the woodcock, where it lies attached to the villi by its suckers and crown of some sixteen rostellar hooks. The mechanism of the rostellum is interesting, because from a position withdrawn within the scolex with the hooks applied closely to it, like the ribs of a closed umbrella (Fig. 32i), it can be protruded like a proboscis with the hooks still lying closely applied. After being forced between villi in this condition the hooks can be erected, to use the same analogy, like the opening of an umbrella, and when projecting laterally help to anchor the worm.

The ripe proglottides, shed from the short strobila, are passed in the bird's droppings, disintegrate in the soil and thus come to be swallowed by earthworms. Within the earthworm the egg-shell and embryophore are digested and the oncosphere bores through the wall of the gut and eventually reaches the coelom where further development takes place. Unlike the condition in *Taenia pisiformis* where the oncosphere gave rise to a single *cysticercus*, in *P. paradoxa* the oncosphere enlarges, becomes hollow, and a series of thickenings appear on the wall, each one of which develops into a detached *cysticercoid*. The sphere, with contained cysticercoids, is known as a *polycercus* (Fig. 32h and Plate 9), and large numbers of these are found in infested earthworms where they are more numerous in the posterior segments. When an infested earthworm is eaten by a woodcock the polycercus is broken down mechanically or by the digestive juices and the cysticercoids are liberated, the digestive juices also causing the eversion of the scolex and the digestion of all parts other than the scolex and neck. From the latter the short strobila develops. This is a case of multiplication in an intermediate host. There are, however, other instances of a different nature where the oncosphere gives rise to new individuals, and these may now be considered.

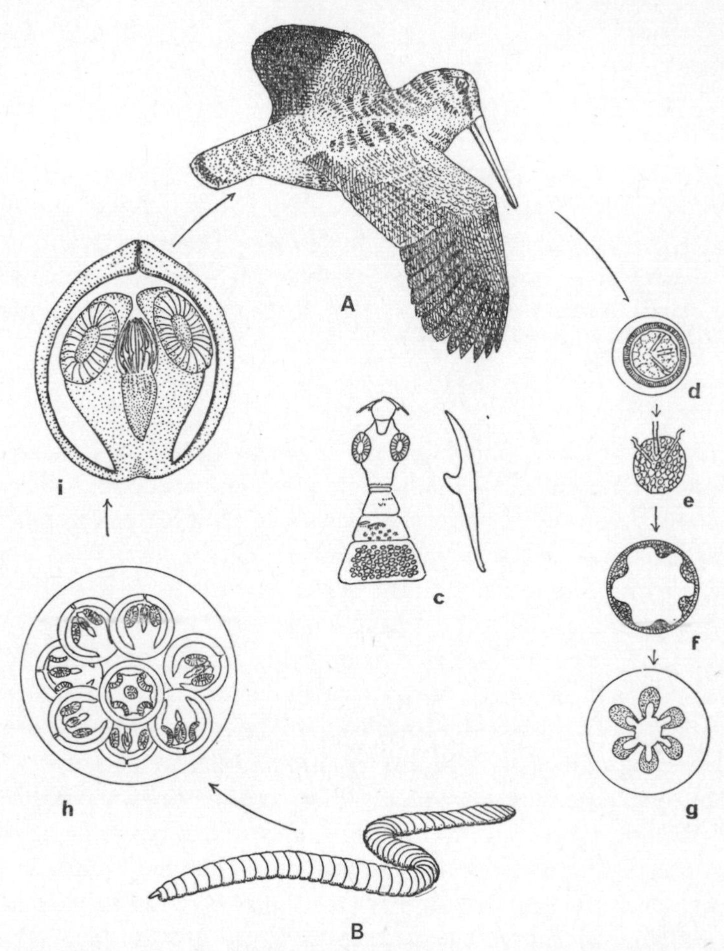

FIGURE 32 Life-cycle of *Paricterotaenia paradoxa* (Rudolphi). The definitive host is a charidriiform bird such as a woodcock, *Scolopax rusticola* L. (A) which feeds on earthworms, and one of these, *Allolobophora terrestris* (Savigny) (B) is the intermediate host. c, complete adult tapeworm and rostellar hook; d, egg with contained oncosphere which is swallowed by the earthworm; e, hexacanth larva, freed from its embryophore, in gut of earthworm; f, g and h, stages in the development of the polycercus with its contained cysticercoids within the coelom of the earthworm; i, single cysticercoid, scolex of which becomes everted in the intestine of the woodcock.

VEGETATIVE OR ASEXUAL REPRODUCTION OF LARVAL STAGES

Sparganum is the name applied to the plerocercoid of some species of *Diphyllobothrium* in which the strobila stage is found in carnivores such as dogs or cats and the plerocercoid normally in frogs or reptiles, but occasionally the plerocercoid may become established in mammals. It is then found in various internal organs where it may attain a considerable size but without, however, segmenting to form a definite strobila or becoming sexually mature. Some cases have been reported of sparganum producing buds which separate from the mother sparganum and develop into new plerocercoids. One may note, by the way, that human sparganosis occurs in some countries, and there is one such condition in Indo-China where the plerocercoids are found in the vicinity of the eye. This, however, is attributable to the ethnic custom of treating affections of the eye by skinning a frog and applying the inside of the skin to the eye, or otherwise applying the tissues of the frog to the eye, comparable probably with the custom in some countries of applying a beefsteak to a black eye. If the frog harbours plerocercoids these may be transferred to man and become re-encapsulated near the eye.

Larval budding

Reproduction by budding is rare in some cestodes but usual in others, and one can distinguish external and internal budding, or exogenous and endogenous budding.

Exogenous budding

Exogenous budding occurs in some species of *Taenia* and a classical example is that of *Taenia crassiceps* (Zeder), a parasite of the fox, with mice or hamsters or even a mole as intermediate host, where the cysticercus is known as *Cysticercus longicollis*. The oncosphere of *T. crassiceps*, ingested by a mouse, makes its way to the subcutaneous connective tissue, and there the bladderworm gives rise to buds which are constricted off and become cysticerci each with a scolex. These in turn may also bud off new cysticerci and there seems to be no limit to the number so formed, for Baer has reported that cysticerci inoculated into the body-cavity of mice and 'subcultured' from time to time continued to reproduce in this manner for more than eight years. External budding has also been seen in cysticercoids where the oncosphere gives rise to an undifferentiated mass which forms a number of external buds, like a small bunch of grapes, and each bud develops into a cysticercoid.

Endogenous budding—coenurus

Endogenous budding may occur rarely in some species, where one or more scolices are formed on the wall of the bladder, one case being the bladder-worm *Cysticercus tenuicollis* found in the mesenteries of sheep and goats. It is the rule, however, in a number of species, where the bladder-worm stages are known as a *coenurus* or as a *hydatid*. Two cosmopolitan cestodes have a coenurus in an intermediate host, *Multiceps multiceps* (Leske) and *Multiceps serialis* (Gervais). In each of these the definitive hosts are carnivores such as dogs, foxes or jackals, but the intermediate host of *M. multiceps* is normally an ungulate, more commonly a sheep, the coenurus being found in the brain. The coenurus of *M. serialis* is found in a rabbit or hare, a common site being the axilla. In each of these the coenurus may have several hundreds of scolices on the wall of the bladder, each scolex having the potentiality of forming a strobilate worm in the definitive host (Plate 7).

Hydatid

The larva of a number of minute taeniids exhibits endogenous budding. An almost cosmopolitan example is *Echinococcus granulosus* (Batsch) which is found in dogs and cats in most parts of the world. It measures from 4 to 6 mm in length and comprises some three or four segments of which only the terminal one is gravid at any particular time (Fig. 33). Compared with the tapeworm the larva is enormous, for it frequently reaches a diameter of 10 cm, though cysts as large as a rugby football are known, and one with a diameter of half a metre has been recorded. Hydatid cysts may develop in a very great variety of mammals including man, though commonly in herbivores, and the usual site is the liver. After arriving in this organ by way of the portal vein, the hexacanth larva increases in size and an adventitious membrane derived from the tissues of the host is formed round it. The wall of the cyst, apart from this outer membrane, becomes thick showing a concentric stratification of layers, and internally it is lined with a germinal membrane or proliferating layer similar to that found within a coenurus (Fig. 33E). The cyst contains a hydatid fluid into which pass the products of the germinal membrane, which can be seen in a number of stages of development. These include scolices that are budded from the germinal membrane, which remain attached for a time and then separate to float in the hydatid fluid. Many of these scolices degenerate and undergo vesiculation to become daughter cysts, and lined with a proliferating membrane they develop within themselves new scolices and are then known as brood capsules. Daughter cysts may also be formed directly from the proliferating membrane of the original hydatid without the intervening scolex-stage. Each brood capsule usually contains from 10 to 30 scolices, but the number may be many more. The hydatid cyst grows slowly, and gradually there accumulate floating within it, and known as hydatid sand, enormous numbers of scolices. It has been estimated that scolices number 400,000 in every ml of hydatid fluid, so that a small hydatid of only 5 cm diameter will contain 26 million of them. Exogenous budding has been recorded in the case of *E. granulosus*, and this is usually attributed to a small part of the germinal membrane being enclosed in the cyst-wall and, passing to the outside of the cyst, it undergoes vesiculation to become a hydatid like the mother cyst that produced it.

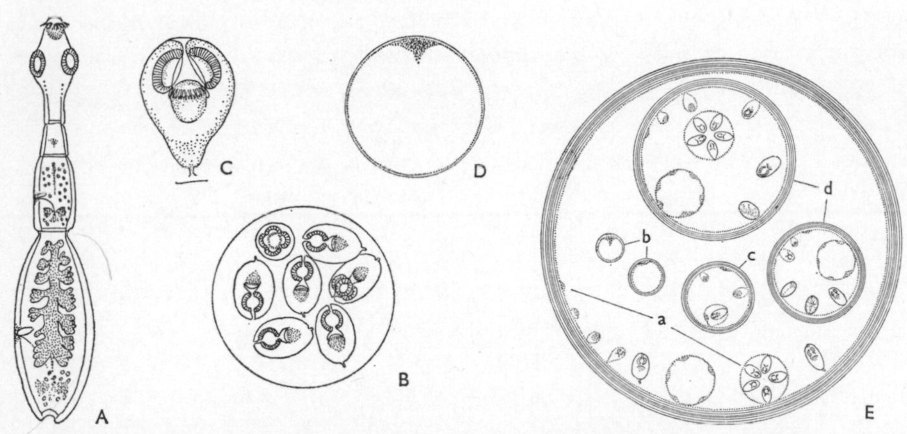

FIGURE 33 *Echinococcus granulosus* (Batsch).
A, strobilate cestode showing three proglottides; young, mature and gravid. B, brood-capsule containing scolices. C, scolex from wall of hydatid cyst or brood-capsule. D, vesiculation of scolex. E, schematic representation of hydatid cyst showing endogenous development of brood-capsules, scolices, and formation of daughter cysts. a, formation of scolex and brood-capsule; b, vesiculation of scolex; c, daughter cyst; d, daughter cysts with scolices and brood-capsules.

Multilocular hydatid

Like *E. granulosus*, *E. multilocularis* is a parasite of carnivores, but the intermediate stage is found in rodents. For a long time this tapeworm was considered to be the same species as *E. granulosus* as the strobilate stages were rather similar, but there is a great difference in the hydatid stages. In *E. multilocularis* the hydatid lacks a laminated external coat, and without the restraint of this coat the germinal membrane grows and extends within the organ infested, usually the liver or lungs, and eventually forms a complex structure, almost like a malignant growth, which is full of cavities, or loculi.

As in the case of *E. granulosus* scolices are formed directly on the germinal membrane as well as in brood capsules (Plate 8).

Polycercus

A type of endogenous budding already considered within the life-cycle of *P. paradoxa* is the formation of a number of cysticercoids from a single oncosphere (Fig. 32h and Plates 9 and 10).

SHORTENED LIFE-CYCLE

The life-cycle may be shortened in ways similar to those seen in Digenea, by eliminating the intermediate host or by neoteny. *Hymenolepis fraterna* (Stiles), sometimes considered as a variety *fraterna* of *Hymenolepis nana* (Siebold), may be taken as an example of the former way in that it shows an alternative life-cycle in which there is but one host. In those species of *Hymenolepis* where the life-cycle is known, the definitive host is a warm-blooded vertebrate and the intermediate host is an invertebrate, usually an insect or a crustacean. In *H. fraterna* the definitive host is a mouse or rat, or some other rodent, with a flea such as the rat-flea *Ceratophyllus fasciatus* Bosc or *Xenopsilla cheopis* (Rothschild) or the meal-worm *Tenebrio molitor* L. as intermediate host. The gravid proglottides of the tapeworm pass out of the rat with its faeces and if these foul the nest of the rat or the grain on which it feeds they may be eaten by the larvae of fleas or of meal-worms. Within the larval insect the oncosphere develops in the usual way into a cysticercoid and when eventually the metamorphosed flea or the meal-worm is swallowed by a rat, infestation is brought about. In the alternative life-cycle the egg hatches while still within the rat and the oncosphere penetrates an intestinal villus in the middle region of the small intestine and within four days it has developed into a cysticercoid. It leaves the villus and as it passes down the gut the scolex becomes evaginated and becomes fixed to the gut-wall where it develops into a normal cestode, thus eliminating the intermediate host. Baer considers that the development in a single host is superseding the development in two hosts and that one may consider the two regions of the intestine, the one where the cysticercoid develops and the one where the tapeworm becomes attached, as intermediate host and definitive host respectively.

There are other instances of the definitive host functioning also as an intermediate host and *Taenia solium* L., the solitary or pig-tapeworm, is one of these, but the case is not strictly parallel. The definitive host of *T. solium* is man, and the intermediate host in which the bladder-worm, known as *Cysticercus cellulosae*, develops is the pig, but as this tapeworm is not strictly monoxenous an alternative intermediate host is also man himself. Man can thus become infested by ingesting eggs of *T. solium* harboured by himself or some other person. Another means is by internal auto-infestation, when by antiperistalsis, or by active movement of shed proglottides, or by both, eggs are introduced into the stomach from the intestine. Here they hatch and the oncospheres settle down in some organ, usually muscle, where they become viable cysticerci. Eventually they become calcified and may remain undetected unless X-ray examination brings their presence to light. If the site of infestation is the brain consequences are much more serious. In the absence of cannibalism man is a dead-end host and the fate of the cysticerci is sealed, a condition unlike that of *H. fraterna*.

The life-cycle can also be shortened by elimination of the adult host, when by neoteny the stage in the intermediate host becomes sexually mature. The classical example of this, that of *Archigetes sieboldi* Leuck., is found in the little red fresh-water oligochaete *Tubifex tubifex* (Müller), in whose coelom it develops to sexual maturity. It is a small worm, some five mm long, and its body shows an anterior broader region and a tail-like posterior region which has six hooks at the end. The anterior region contains a single set of hermaphrodite organs showing a typical pseudophyllidean arrangement. When the cestode becomes sexually mature it escapes from its annelid-host by forcing its way through the body-wall, its disintegrating body liberates the eggs and from these there hatch non-ciliated coracidia which in turn are swallowed by other tubificids. A tapeworm considered to be the same parasite is the fish-cestode *Bialovarium sieboldi* Szidat found in Europe in the tench *Tinca tinca* (L.), and which, like *Archigetes*, is unsegmented.

The general pattern of the life-cycle of Pseudophyllidea is that of *D. latum* (Fig. 30) in which the coracidium becomes parasitic in an invertebrate where it develops into a procercoid, and on being eaten by a fish the procercoid develops into a plerocercoid. *Archigetes* follows this pattern to a limited extent for it becomes a neotenic procercoid in *Tubifex*, still retaining its oncosphere-hooks, and it can continue as a neotenic plerocercoid in a tench, but it lacks the segmented stage in a fish-eating animal such as might be expected. This is not an isolated case of neoteny for there are numerous genera comprising a considerable number of species grouped in the family Caryophyllaeidae, which are all unsegmented forms and presumably neotenic plerocercoids. An interesting and provocative suggestion is that we are dealing here with the descendants of parasites whose original definitive hosts have become extinct but which have managed to survive as neotenic forms. The question arises whether it is possible that these hosts could have been Mesozoic reptiles which did not survive the cataclysm called the Laramide revolution towards the end of the Cretaceous. However, it seems more likely that there is a general tendency towards neoteny in this group and one can arrange the genera in a series showing degrees of neoteny in Caryophyllaeidae. Just how near this can come to a segmented worm in a single host is seen in the parasite of an amphipod, the sand-jumper *Marinogammarus pirloti* Sexton and Spooner. Within the haemocoel of this amphipod there have

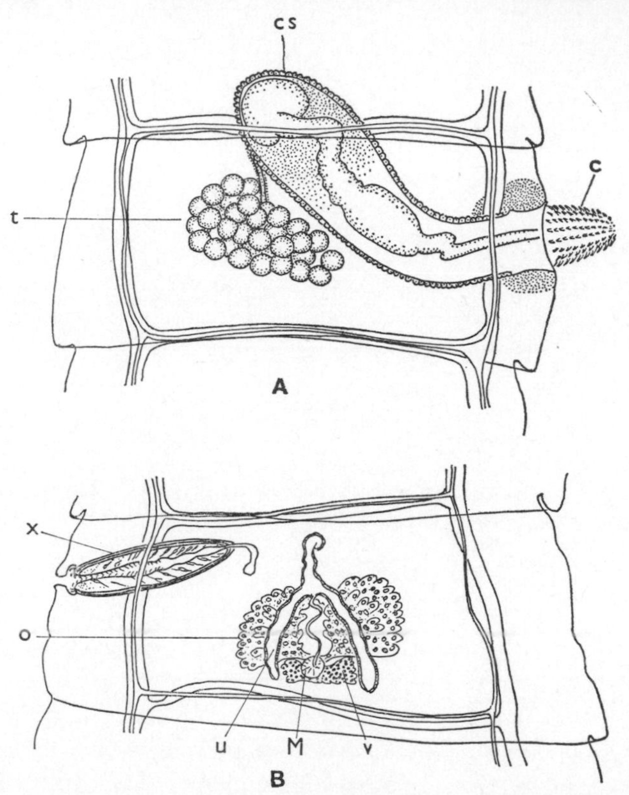

FIGURE 34 The reproductive system of a dioecious cestode *Infula* sp. from the North American dowitcher, *Limnodromus griseus hendersoni* Rowan. A, Proglottis of a male worm showing testes (t), large cirrus-sac (cs) and a partially everted cirrus (c). B, Proglottis of a female worm showing bilobed, fan-shaped ovary (o), vitelline gland (v), Mehlis's gland (M), horseshoe-shaped uterus (u) and a cirrus-sac-like body (x).

been found specimens of *Diplocotyle nylandica* Schneider apparently mature, strobilate, caryophyllaeid cestodes in the uterus of which there were eggs, and this stage has been described as a neotenic form parasitic in an invertebrate. Dr Michael Burt has reported that in the specimens examined by him the eggs are infertile and in fact the testes are not developed, but that the testes do develop when the parasites from the haemocoel of *Marinogammarus* are introduced into a teleostean fish such as a plaice, *Pleuronectes platessa* L.

DIOECISM IN CESTODES

At the beginning of this century Fuhrmann recorded a dioecious cestode, *Dioecocestus paronai*, from a white-faced glossy ibis, *Plegadis guarauna* (L.) of South America. He found associated in this bird separate male and female worms, the former with a double set of reproductive glands and two cirri in each segment, and the latter, larger and more robust, with but a single set. Since then other dioecious species of this genus have been found in grebes. The hosts, ibises (Ciconiiformes) and grebes (Podicipediformes) are by no means closely akin, but both share the same biotope and feed on small fish, insect larvae, crustaceans and other aquatic life. Another dioccious genus, *Infula* Burt, is found in wading birds (Charadriiformes) such as stilts, stone-curlews and dowitchers or red-breasted snipe. In it the sexes are also separate with disparity in size but there is a single set of reproductive organs in the male while the female is characterised by the possession of a peculiar cirrus-like organ (Fig. 34). More recently (1968) Dr Harford Williams has reported another dioecious cestode which is a tetraphyllidean parasitic in an elasmobranch. He has actually observed connection between the male and female worm which confirms that they belong to the same species.

Dioecism is an instance of parallel evolution within the phylum Platyhelminthes for it also appears in Digenea in schistosomes, where the mechanism is genetic and in Didymozoonidea where the determining factors are apparently environmental, but the cause in Tetraphyllidea and Dioecocestidae has not yet been determined.

HOSTS OF CESTODES

With very few exceptions a cestode has more than one host. In some orders, as in Pseudophyllidea, there are generally three hosts, while in Cyclophyllidea there are two. The relation of one host to another is either nutritional where it is that of predator and prey, or ecological where infestation is frequently a matter of chance. The relationship is nutritional as when a dog eats a rabbit, a cat a rat, and a bird an earthworm where the prey is infested with the larva, and this may also be the case when an intermediate host becomes infested with a larva, as when a crustacean eats a coracidium or a fish a crustacean. But the relationship is not that of prey and predator when a rat swallows a mealworm, or an ox ingests a mite or a dog bites at and swallows a dog-flea. In these instances there is chance and association in the same habitat of definitive and intermediate hosts. In the orders containing the more generalised tapeworms such as Tetraphillidea, Pseudophyllidea or Tetrarhynchidea there is a free-living stage in the life-cycle for the coracidium has a short existence in the outside world before it reaches the intermediate host, but in the Cyclophyllidea which are parasites of birds and mammals, the embryo remains within its embryophore during transference from definitive to intermediate host and the larva has no free-living existence.

In the accompanying table which gives the hosts of a number of tapeworms selected from different orders the relationship between definitive and intermediate hosts can be deduced.

TABLE 6

Definitive hosts and intermediate hosts of tapeworms selected from different orders

Tapeworm	Definitive hosts	Intermediate hosts
HAPLOBOTHRIIDEA		
Haplobothrium globuliforme Cooper	fish (Holostei) bow-fin (*Amia calva*)	crustaceans and bony fishes Procercoid in *Cyclops viridis* Plerocercoid in *Ameiurus nebulosus* and *Eupotomus gibbosus*
PSEUDOPHYLLIDEA		
Diphyllobothrium latum (L.)	bear, dog, man	crustaceans and bony fishes Procercoid in *Cyclops strenuus*, *Diaptomus oregonensis*, etc. Plerocercoid in *Salmo*, *Esox*, etc.
Archigetes sieboldi Szidat	oligochaete worm or fish worm—(*Tubifex tubifex*) fish—tench (*Tinca tinca*)	oligochaete worm Procercoid in *Tubifex tubifex* in which it may become mature
Diplocotyle nylandica Schneider	fish (Teleostei) plaice (*Pleuronectes platessa*)	amphipod crustacean strobilate and all but sexually mature form in *Marinogammarus pirloti*
TETRARHYNCHIDEA		
Grillotia erinaceus (v. Beneden)	fish (Elasmobranchii) skate (*Raja* spp.)	crustaceans and bony fishes Procercoid in *Pseudocalanus*, etc. Plerocercoid in *Gadus*, *Clupea*, etc.
TETRAPHYLLIDEA		
Anthobothrium cornucopia v. Beneden	fish (Elasmobranchii) sharks, dogfishes, rays (*Carcharias*, *Lamna*, *Raja*, *Trygon*, etc.)	bony fishes Procercoid probably in a crustacean Plerocercoid in a bony fish
ICHTHYOTAENIIDEA		
Ophiotaenia perspicua La Rue	Reptile—snake diamond-backed water-snake (*Natrix rhombifera*, etc.)	crustaceans and amphibians Procercoid in *Cyclops vulgaris*, etc. Plerocercoid in tadpoles and adults of green frog (*Rana clamitans*), etc.

TABLE 6—*continued*

Tapeworm	Definitive host	Intermediate hosts
CYCLOPHYLLIDEA		
Taenia solium L.	man	pig Cysticercus in *Sus scrofa*
Taenia saginata Goeze	man	oxen Cysticercus in *Bos taurus, B. indicus,* etc.
Taenia pisiformis (Bloch)	dog	hares and rabbits (Lagomorpha) Coenurus in *Lopus, Oryctolagus,* etc.
Taenia crassiceps Zeder	foxes *Vulpes, Alopex,* etc.	rodents (Myomorpha, Sciuromorpha) Cysticercus in mice (*Mus, Citellus,* etc.) squirrel (*Sciurus*)
Taenia taeniaeformis (Batsch)	cats, stoats, pine-marten Felidae, Mustelidae	rodents (Myomorpha) Strobilocercus in rats and mice
Echinococcus granulosus Batsch	dog	sheep and oxen (Bovidae) Hydatid in *Ovis aries, Bos taurus, B. indicus,* etc.
Echinococcus multilocularis Leuckart	dog	rodents (Myomorpha) Multilocular Hydatid in field-mice, rats, etc. (*Citellus, Microtus, Mus,* etc.)
Ophryocotyle insignis Lönnberg	bird—oyster-catcher *Haematopus ostralegus*	limpet (Gastropoda) Cysticercoid in *Patella vulgata*
Paricterotaenia paradoxa (Rudolphi)	birds (Charadriiformes) woodcock (*Scolopax*), etc.	earthworm (Oligochaeta) Polycercus in *Allolobophora terrestris*
Dipylidium caninum (L.)	dog	fleas (Aphaniptera) Cysticercoid in dog-flea (*Ctenocephalus*), etc.
Hymenolepis diminuta (Rudolphi)	rodents (Myomorpha) *Rattus, Mus, Citellus, Apodemus, Arvicola etc.*	mealworms and fleas (Insecta) Cysticercoid in *Tenebrio, Tribolium* etc. *Ceratophyllus, Xenopsilla* etc.
Raillietina tetragona (Molin)	birds (Galliformes) *Gallus, Numidus, Lagopus,* etc.	flies and ants (Insecta) Cysticercoid in *Musca, Pheidole,* etc.
Moniezia expansa (Rudolphi)	ruminants (Bovidae) *Ovis, Bos, Capra, Ibex, Antilope,* also *Rangifer*	mites (Acarina) Cysticercoid in oribatid and tyroglyphid mites (*Galumna, Oribatula, Scheloribates,* etc.)

11 *Host-Parasite Relations*

To every action there is always
an equal and contrary reaction.
Newton's Third Law of Motion

The great impetus to the study of the effect of parasites on their hosts arose from Pasteur's work in the field that was to become Bacteriology because of its direct application to man and his domestic animals. The study of the reactions of the human body to the presence of bacteria has been extended to the reactions produced by parasites other than bacteria and to the fundamental properties of living matter.

IMMUNITY AND PREMUNITION

Some parasites never become established in one particular host although there is no barrier to their establishment in another and one labels the condition as *natural immunity*, which of course describes the phenomenon but does not explain it. Most organisms, apparently, have this property to some degree, but it can be developed or increased as a result of infestation by a parasite, or in other ways to be described, when it becomes an *acquired immunity*. In investigating this mechanism of the defence of the living body, one looks first of all at the functioning of the body in dealing with damaged tissues, by the removal of the damaged cells and the regeneration of new tissue. Cells that are effete are broken up and the resulting materials are re-synthetised. Foreign substances may be surrounded by a growth of connective tissue and encapsulated, and foreign proteins are broken down in the liver where deamination takes place.

In the lymph-glands of the body of a vertebrate certain cells known as lymphocytes are formed and these circulate through the lymphatics and in the blood where they are often referred to as plasma-cells. Such cells are also produced in the spleen and in the bone-marrow, and it is in these lymphocytes that one finds the mechanism against, and reaction to, the presence of foreign substances in the body. Lymphocytes are frequently seen in large numbers in the tissues surrounding parasites.

Lymphocytes react to foreign substances, such as proteins, or the presence of a parasitic organism, or it may be to the substances produced by such organism in the manufacture of *antibodies*. An antibody is a complex protein, globulin in nature, which can be identified by its reaction, *in vitro*, with the substance known as an *antigen* that provoked its production. The whole story of the origin of the particular cells which produce antibodies, of the manner in which these cells respond to an antigen and the mechanism of the elaboration of the antibody is not yet unfolded in its entirety. There is, however, a general consensus of opinion that one of the sites is in the lymph glands, and in these glands the particular lymphocytes, the plasma-cells, are formed, though some would trace the ancestry of these cells farther back to the tissues of the thymus gland. This gland, lying just in front of the heart, which is comparatively large in young animals but becomes reduced in size as animals grow up, appears to be involved with the synthesis of DNA, deoxyribonucleic acid, whose long chain-molecule has a genetically significant, information-carrying role. The thymus is also associated in some way with the ability to produce antibodies. That plasma-cells respond to the presence of antigens is acknowledged, but the mechanism of the response is still a matter of conjecture. In the main there are two opposing views with regard to this. One is that these cells, acting something after the manner of a templet, mould the antibody they produce against the invading antigen, and thus formed the antibody combines with the antigen and inactivates it. The other view is that the cells which produce the antibody, the plasma-cells, are endowed genetically with the ability to 'recognise' a particular foreign substance and are provided with the material which enables them to multiply rapidly and produce the specific antibodies that will combine with the antigens that they 'recognise'. One's mind boggles at the idea that a single small cell could be endowed genetically with the ability to recognise and respond to one particular antigen out of the thousands that it might encounter, but it may not be too far-fetched to compare this ability with that of being able to recognise a particular smell or taste. In taste, for example, it is apparently the shape of the molecule that is responsible for the particular stimulus and there are not many more than half a dozen fundamental molecular shapes involved in the vast numbers of flavours that man can appreciate. The plasma-cells may, in like manner, recognise a limited number of molecular shapes, a view that has something in common with the opposing 'templet' and 'selective' theories.

The production of antibodies has been traced in the laboratory in single plasma-cells kept in microdroplets. The stimulus of antigen results in an increase in ribosomes in the cell, the microscopic cytoplasmic bodies involved in the production of RNA, ribonucleic acid, the substance which directs protein synthesis. Antibody which is a protein, γ-globulin in nature, is produced and passes into the surrounding medium. Electron-microscopic techniques show that this is accompanied by the appearance in the cell of an endoplasmic reticulum, such as is found in secreting cells.

The reaction of antibody, present in a soluble form in the blood serum of the animal that produces it, with the antigen responsible for its production, may be made manifest in a number of ways. From these there have been developed various laboratory techniques which are qualitative and quantitative in nature, and some of these are the following.

PRECIPITATION

A combination of antigen and antibody in the form of a precipitate results when a soluble antigen is added to blood-serum containing the antibody which it has produced, the serum being known as *antiserum*. In terms of actual experiment one can use as an antigen the homogenised and filtered extract of a fluke, and inject this at intervals over a number of weeks into a rabbit. If, later, one mixes the serum of this rabbit, that is the antiserum, with the antigen, that is the extract of the fluke, and allows the mixture to stand in an incubator for half an hour, an insoluble precipitate separates out and appears as a cloudiness. Alternatively, one can demonstrate the presence of a parasite such as a hydatid in an animal by adding to the blood-serum of this animal a small quantity of hydatid fluid from a cyst, when a precipitate appears in the mixture. One can use this test quantitatively using different dilutions of antigen, and by measuring the degree of cloudiness obtained in each case obtain an estimate of the concentration or titre of the antibody present in the antiserum.

A more sophisticated technique, but one using the same fundamental principle, is that of allowing antiserum and antigen to diffuse through a gel, such as one of agar-agar when a precipitate is formed where the two diffusing substances meet. This is the plate technique of Ouchterlony where the gel is a thin plate with wells cut in it, usually a central one with a few arranged equidistantly around it. One can put antiserum in the central well and various antigens which are to be tested in the surrounding wells when the appearance of a number of lines of precipitate between the various antigens and the antiserum indicates the presence of different components in a complex antibody. This method can also be used to compare antisera produced in response to different parasites, or in response to various stages in the life-cycle of one parasite.

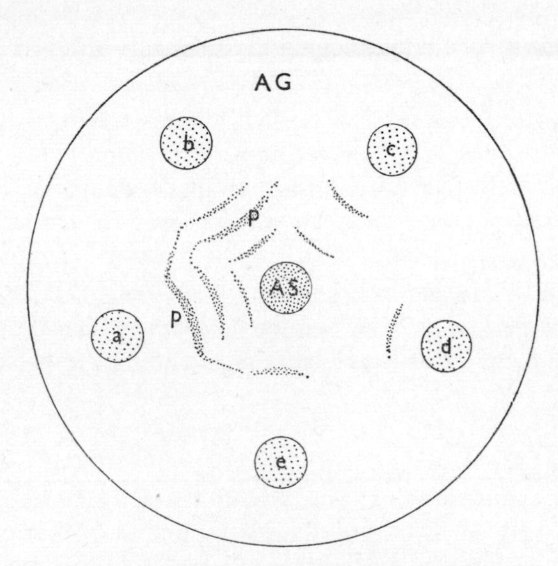

FIGURE 35 Plate technique of Ouchterlony.
Precipitative reactions of antigens and antisera in an agar-gell diffusion plate (AG). a, b, c, d and e are wells cut out of the agar plate containing different antigens while AS is a well containing antisera. Precipitates (p) are formed in bands where the diffusing substances from the antigens react with the diffusing antisera. AS might be blood serum of a rabbit containing antisera to a particular fluke, with a, b, c, d and e antigens such as the test antigen used in obtaining the antisera (a), and larval antigens from miracidia, rediae and cercariae, or from other flukes.

AGGLUTINATION

When the antigen is in the form of cells, bacteria, blood-corpuscles or protozoa, the reaction to the homologous or specific antibody is a clumping or agglutination of these particles. An extension of this phenomenon is the use of various inert particles which have been treated in order to coat them with the antigen to be tested.

HYPERSENSITIVITY

The presence of parasites in an animal can sometimes be diagnosed by the reaction of the skin to an extract of the suspected parasite or even to an aqueous extract of the dried and powdered parasite. This can be carried out by placing a drop of the extract on a small area of scarified skin or injecting a small quantity into the skin itself when a positive reaction is the appearance of a weal at the place of application in about a quarter of an hour. This is just an application of the technique commonly used in testing for an allergy.

ANAPHYLAXIS

A manifestation of hypersensitivity is sometimes seen when the reaction of antibody to antigen occurs with great violence, as in the reaction of an animal with a hydatid cyst to hydatid fluid, should the cyst be ruptured. In this instance the reaction occurs in the tissues of the body and not in the bloodstream, and the result is called anaphylactic shock, which is accompanied by lowering of the blood pressure, collapse of the animal and sometimes death.

ELECTROPHORESIS

The separation and identification of the different substances present in antisera can be carried out by electrophoresis. The essence of this method is passing an electric current across a sheet of paper, or some suitable supporting medium such as agar or cellulose acetate, with one edge of the paper dipping in the antiserum to be tested. The various proteins present in the antiserum are separated according to their electric charges as they are carried across the paper. In immuno-electrophoresis, an extension of this method, various constituents of the antigen are separated across a sheet of agar by electrophoresis, and when antiserum is added, on strips of filter-paper placed along each edge of the agar sheet, the reaction of antigen and antiserum can be observed as separate precipitates.

FLUORESCENT ANTIBODY TEST

Some dyes have the property of absorbing light of one wavelength and emitting light of another, and one of these dyes is fluorescin isothiocyanate which has also the property of combining with certain proteins and which can, accordingly, be used to combine with antibodies. If a dye-conjugated antibody is applied to the homologous antigen it combines with the antigen which itself becomes fluorescent. This serological staining technique has application in the study and identification of very minute parasites such as blood-protozoa—piroplasms and plasmodia—when recognition of various stages in the life of these parasites can be made under the microscope using ultra-violet light, and a quantitative measure of antibody present can be made.

COMPLEMENT FIXATION

An adaptation of the test for syphilitic infection known as the Wassermann test has been used to determine the nature of parasitic infestation. The basic principle of this test lies in a property of normal blood to lyse or destroy foreign organic bodies such as bacteria, protozoa or blood-corpuscles. This property is due to constituents of normal blood called *complement*, which however is thermolabile and becomes inactive if heated to 55°C. Complement is present in the blood of different species of animals in differing proportions, but since guinea-pig serum contains a relatively high amount and as this animal is readily available, it is usually used in the test.

In brief outline the procedure involves the use of a *haemolytic serum* for blood-corpuscles of sheep. This is usually rabbit's serum, and it can be purchased ready for use, or prepared by injecting sheep's blood-corpuscles, washed in saline, into a rabbit over a period of weeks and then separating the rabbit's blood serum from its blood. In the presence of complement this haemolytic serum will lyse sheep's blood-corpuscles. Thus

corpuscles + haemolytic serum + complement → lysis of corpuscles

To determine whether an animal is infested with a parasite, such as a sheep with liver-fluke, serum of the sheep suspected of being parasitised is heated to 55°C in order to destroy its thermolabile complement, and to this is added the *test antigen* in the form of a triturated and filtered extract of the fluke suspected and also some fresh serum of guinea-pig containing, as it does, active haemolytic complement. If the test antigen is homologous with the antiserum these combine together and with the guinea-pig's complement so that there is no free complement in the system. In the absence then of free complement the addition of *sensitised sheep's corpuscles*, that is the blood-corpuscles of sheep washed in saline and treated with haemolytic serum, does not produce a lysis of the corpuscles. On the other hand, if there is no reaction between the test antigen and the suspected sheep's antiserum there will be free, active, guinea-pig haemolytic complement present which will cause the corpuscles to be lysed. Thus a negative result, absence of lysis, means a positive infestation, and a positive lysis indicates a negative result as regards the fluke in question.

There are several theories regarding the nature and mode of production of antibodies, but although these still remain unknown the presence of such substances in the blood-stream is fully established.

The intensity of the reaction of the hosts depends on the site of the parasite. It is obvious that the reaction to a parasite located in the tissues of the body, in direct contact with lymph and lymphocytes, will be greater than the reaction to one lying in the lumen of the intestine, where the antigen will consist of such substances produced by the parasite which are absorbed through the wall of the gut. But even in such cases antibodies are frequently adequate in restricting the site of infestation, in limiting the reproduction of the parasite, in preventing further infestation and even producing expulsion of the parasite. The reaction to a single gut-parasite is seen in the case of some of the larger cestodes, such as *Taenia solium* or *T. saginata* where a single individual is sufficient to inhibit the establishment of more parasites of the same kind, that is it prevents super-infestation. This reaction is more pronounced where the stage in the life-cycle is the bladder-worm stage, when the presence of a single cysticercus in the liver or mesenteries elicits a protective immunity. Attempts have been made, and with some measure of success, to immunise animals against cestode infestation by injecting, as antigens, either extracts of the parasite in question or of its eggs after subjecting them to X-rays, or of its artificially hatched oncospheres. In addition to protection against challenging doses of the homologous parasite there was also some degree of protection against other, but different parasites, indicating the very complex nature of the antigens and the antibodies they elicit. A further result was that complete immunity was only obtained using living organisms as antigens thus indicating the qualitative production of antibodies.

IMMUNISATION

Some attempts to immunise animals against infestation by cestodes have been carried out by injecting them with the blood-serum of infested animals and then 'challenging' the immunity by attempting to parasitise these animals. This has been successful in the case of bladder-worms, such as *Cysticercus pisiformis* of the rabbit, *Strobilocercus taeniaeformis* of the rat, and of the tapeworm *Hymenolepis fraterna*, where strobila and cysticercoid are found in the same host. It has also been established that immunity varies during the course of an infestation, that the immunity first acquired may differ in quality from that produced when the parasite is well established. This is a further indication of the complex nature of the reaction between host and parasite and it is the elucidation of phenomena such as these that must lead to the control, or eradication, of parasites from man and his domestic animals.

In the case of parasitic flukes living in man himself, one of the schistosomes, *Schistosoma mansoni*, has attracted the attention of many workers, one of the reasons being that it has an adaptation that enables it to live in a number of different hosts and accordingly is a useful laboratory animal. Although it has been possible to induce an absolute immunity to this parasite in some laboratory animals, success in eliciting immunity through vaccination has not yet been accomplished.

Many parasites are known to persist in the blood of their hosts for very long periods and among these are trypanosomes and malarial parasites among Protozoa, filaria among nematodes and schistosomes among Platyhelminthes. Recent work by Dr S. R. Smithers and his colleagues has thrown light on this phenomenon as it relates to schistosomes. They found that when schistosomes which had been reared in one animal were removed from its portal vein and transferred to that of a different animal, the metabolism of the schistosomes was seriously affected at first but within a short period the transferred schistosomes were fully adapted to their new environment. In effect, schistosomes reared in a mouse could overcome the immune defences and become fully established in a monkey. But if the monkey had been previously immunised against mouse-tissue, the schistosomes transferred from the mouse soon died. The conclusion is that schistosomes grown in a mouse have mouse-antigens associated with them, and further investigation shows that the site of the antigen is in the tegument of the schistosome. It remains to be seen whether similar reactions obtain with other parasites which live within the blood-stream or on the blood of their hosts.

Premunition

In the case of some parasites, immunity against a new infestation lasts only so long as parasites are present in the body, and after parasites are lost, either naturally or by expulsion, re-infestation is possible almost immediately. This phenomenon is known as *premunition*. It appears that the conditions brought about by the initial infestation create an atmosphere inimical to other parasites of the same species. The mechanism is undoubtedly not entirely due to the presence in the host of the homologous antibody and it is possible that a crowding-effect is involved. It is possible that the initial parasites have developed some defence mechanism against the antibodies of the host and these act against the newcomers.

Self-cure

The phenomenon, where there is a mass expulsion of gut-parasites, was labelled self-cure by Stoll. It was observed in some infestations by nematodes where a super-infestation brought about a mass expulsion of the parasites first established. One explanation is that the reaction is in the form of a local anaphylaxis. So far the phenomenon has not been recorded in flatworms.

Other factors

The factors preventing or controlling the establishment of a parasite in a host vary from one kind of parasite to another and frequently involve reactions other than serological ones. This is so in the case of ectoparasites where the reaction may be a simple reflex such as scratching, or it may not depart from the customary toilet habits of the animal such as licking, preening, combing or rubbing the body against some object. A dog, biting at the place where the dog-flea irritates, may manage to bite it and get rid of it, but this habit is also involved in the infestation of the dog with the tapeworm *Dipylidium caninum* (L.) whose intermediate host is the dog-flea. Scratching, preening, biting and rubbing are simple behavioural reactions which can be observed in birds and mammals and which may be related to ectoparasitic insects and arachnids, but there are others of a more complicated nature. One of these is 'anting' where the bird picks up ants with its beak and pushes them amongst its feathers. Whether ants, as predators, can cope with organisms as large as Mallophaga is a moot point, but the author has seen them picking up and eating ectoparasitic avian mites. Salmon, fresh-run from the sea, have their copepod parasites known as gill-maggots *Lernaeopoda salmonea* (L.) and sea-lice *Lepeophtheirus salmonis* (Kröyer), but these do not survive long after the fish enters fresh water.

Parasites will in due course die, and in the case of gut parasites be eliminated in the faeces. But some parasites become immured by the reaction of the host in forming a capsule of connective tissue around them, and thus encapsulated they are put out of circulation. They may remain in this condition if they are larval stages until their host becomes the prey of another animal, though they sometimes die and degenerate, as can be seen in some hydatid cysts and in cystercerci. Some cystercercoids of tapeworms of certain marine fishes become surrounded by concentric layers of calcite in their molluscan hosts, and thus become transformed into the nacreous sepulchres called pearls.

HOST SPECIFICITY AND PHYLOGENY

It is well-nigh a hundred years since Krabbe, in 1869, on the evidence then available, found that for each group of birds there was a particular fauna of cestodes, and that one could envisage the evolution of birds and of their tape-worms proceeding side by side. Since then a study of the cestode-fauna of other classes of vertebrates has brought to light the fact that in these groups too there is a certain degree of specificity. The trail having been blazed by Krabbe for a study of host specificity in cestodes, it fell to other workers to examine the relationships existing between hosts and their parasites in other parasitic groups. As a result a truly formidable amount of information on host specificity has been amassed for every known group of parasites, but in particular the Protozoa, Platyhelminthes, nematodes, insects and arachnids. It would be quite impossible, in an essay of this nature, to deal with any degree of thoroughness with the host and parasite relationships of even one of these groups, so it must suffice to select a few instances to illustrate some of the phenomena and problems that this kind of study has brought to light.

In the first place, as Baer so wisely insists, one has to be very sure of one's facts. There are species known from a single specimen, recorded but once, it may be fifty years or more ago, and then not seen again; there are others now known to be composite species, where parts of different cestodes found together in one host have been mounted on one slide and considered as one species; and there are others where the host is incorrectly identified or there has been a mix-up of labels. Another danger is in the scientific literature itself, so enormous and so dispersed, that some workers, unable to see original specimens or read original documents, rely on abstracts and compilations, and of course, once errors appear in these, as they must inevitably do, they are repeated and come to be regarded as incontrovertible data. At the same time when apparent facts do not agree with theory they must be faced until proved wrong. A case in point, which has just come to light, is that of the only known monogenean parasite on a mammal, *Oculotrema hippopotami*, described by Stunkard in 1924 from five specimens which were not collected by him but were believed to have been collected by Looss from the eyes of a hippopotamus in the Cairo Zoological Gardens. Various workers have expressed doubt as to the identity of the host of this material, and naturally so, as the parasite shows close affinities to the Polystomatidae, a family well known from *Polystoma* of the bladder of the common frog, but other members of this family are also found in the mouth, pharynx, oesophagus and urinary bladder of other amphibia as well as in reptiles. Recently, however, numerous specimens of the same species have been reported from Uganda by Dr J. P. Thurston and there also on the eyes of hippopotamuses. The implications are that this worm is either shared with an amphibian or a tortoise, due to the ecological association of these animals, or that the parasite was acquired through such an association and is adapted to live on the eye of this relatively new host. The presence of small, immature forms together with mature forms lends support to the latter view, for this ectoparasite is fully adapted to its habitat around the nictitating membrane and under the eyelids of the hippopotamus.

The geological history of vertebrates is well established although more so in some classes than in others, and this kind of information has been used as criteria in discussions on the origin of parasitism. The fact that Cestodaria are found as gyrocotyleans in Holocephali and as amphilinideans in fresh-water Teleostomi has been used as an argument that the ancestral cestodarians became parasitic in the group providing the common ancestors of the chimaeras and bony fishes. On the other hand the presence of a cestodarian in a fresh-water tortoise is accepted as due to ecological association. And here again, as in the case of *Oculotrema*, one has to distinguish between phylogenetic relationships and ecological association, which is one of the problems that arise in determining host specificity. Frequently it is no problem at all, for one finds many allied families, genera and species of parasites infesting allied families, genera and species of hosts.

It is worth while considering again, but from the point of view of host specificity, some of the factors with which a parasite has to contend. The habitat of a parasite, whether internal or external, has an environment in which different mechanical, physical and chemical factors play a part. In the case of an ectoparasite there is the nature of the surface of the body to which it must adhere. In the parasite natural selection has resulted in the development of various organs of attachment or adhesion, frequently very specialised, which enable it to remain attached to its host. The concept of natural selection is deliberately reintroduced to emphasise the fact that the rapport that exists between parasite and host is not a teleological one, that the parasite is not adapted in order to live on a particular host, but is pre-adapted, that the various bodily structures that have developed are made use of after they have appeared; in other words, that structure precedes function. The shape of the parasite is also related to its environment as has been seen in the case of Monogenea, Mallophaga and Anoplura. Organs of attachment, comparable in so far as they also show a high degree of specialisation in relation to the surfaces to which they are attached, are also characters of endoparasites. Thus, a highly specialised parasite on one host might not be able to establish a bridgehead on another even if it did effect a landing. In relation to physical and chemical factors the parasite also shows a high degree of specialisation, whether for the micro-climate or the food available. There are other factors contributed by the host that also affect the relationship such as natural resistance, premunition and sensitisation. Accordingly, a mutual relationship, established between a particular kind of parasite and a particular kind of host, through millions of years, is unlikely to be replicated in other parasites and their hosts. On *a priori* grounds therefore, there must exist a certain host specificity of phylogenetic origin. A few examples will illustrate this. The flattened sucking lice and compressed fleas of mammals, the long, thin and flattened biting lice of the wings of birds, the claws of whale lice and parasitic isopods and other ectoparasites, and the suckers, hooks, clamps and so on of other ectoparasitic groups and the very elaborate organs of attachment of flukes and cestodes, are all very specialised structures. Even where there is no specialised organ of attachment as in many protozoa, one frequently finds that one protozoan will not become established in another host. There are parasitic Protozoa so similar in appearance that it is not possible to tell one from another on purely morphological grounds, but one will thrive in one particular host and will not parasitise another and *vice versa*. The differences are physiological. It is here that the chemical and physical differences of the micro-climates of the host, including the reactions of the host, play their part.

The opportunities for transference from one host to another are greater where the same biotope is shared by different animals, and it is possible that there are many abortive transfers before a parasite can become established in a new host, but such transfers do occur and then one finds an ecological specificity and not a phylogenetic one. Euzet has shown that many parasites of the cestode group Phyllobothrioidea are specific for various selachians—sharks, dogfishes, rays, torpedoes—but that those in the benthic species of selachians are less host-specific.

The monogenean parasites of the Clupeidae (herring family) and Scombridae (mackerel family) belong to the Mazocraëidae and in this family the genera *Mazocraës* Hermann, *Mazocraëoides* Price and *Neomazocraës* Price are found on the former and *Octostoma* Kuhn and *Pseudacanthocotyla* Yamaguti on the latter. Now herrings belong to the order Isospondyli while mackerel are Percomorphi, which are by no means closely related, but mackerel and herrings are both pelagic, gregarious fishes and as mackerel are predators of young fry, particularly of herrings, the consanguinity of their parasites is not dependent on phylogenetic relationship but on an ecological association and contact. This is shown thus by Bychowsky.

Genera of the Mazocraëidae and their hosts

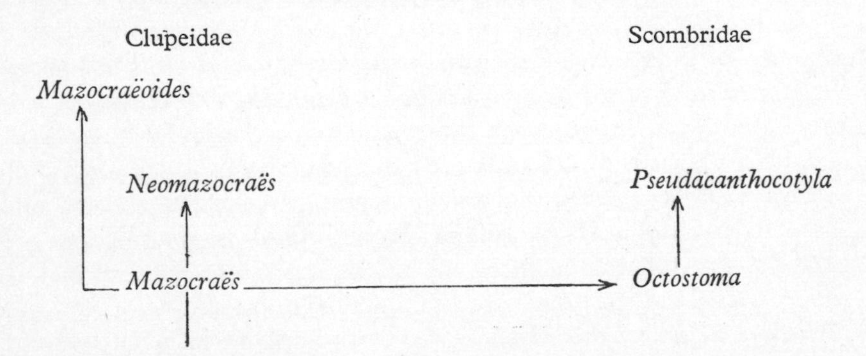

Another type of specificity is site specificity, where there is a large degree of divergence with regard to host but a very strict specificity with regard to location. Such is the case of the strongyloid genus *Syngamus* of which the gape-worm *Syngamus trachea* (Montagu) of the domestic fowl is well known as a parasite of the trachea. This same species is cosmopolitan and infests the respiratory tract of birds of such diverse orders as those belonging to the Passeriformes (perching birds), Anseriformes (water-fowl), Pelicaniformes (cormorants), Otidiformes (owls) and Ciconiiformes (storks), while other species are also found in Sphenisciformes (penguins), Charadriiformes (waders), Accipitriformes (hawks), Gaviiformes (divers) and a cassowary. A still greater variety of hosts is recorded for *Capillaria* Zeder, for this genus parasitises all classes of vertebrates and over a hundred species have been recorded from birds alone. It is noteworthy that here too there is a site specificity with some of the species, for one can instance the case of two different species, both of which are found in two different hosts, but one is an invariable parasite of the oesophagus and the other of the intestine.

There are found, in different parts of the world, birds that are a puzzle to the taxonomist, birds that will not fit nicely into a man-made system of classification, and the evidence of host specificity has been invoked in an attempt to determine their relationships. Among these birds are the kiwis of New Zealand, the cosmotropical flamingos and the seriemas of the pampas of Brazil. The kiwis are flightless birds but their anatomy, apart from the reduction of the wings, is unlike that of the emus and cassowaries. They have a single genus of Mallophaga, *Rallicola*, and a different species of this insect infests each of the three species of kiwis. As these mallophagans are related to species that infest the rails, this corroborates some of the anatomical evidence that these birds are flightless rails.

The case of the flamingos is rather different. Flamingos resemble storks in their long legs and ducks in the general characters of the bill, their webbed toes and in the minute structure of the feathers. The nesting habit is also like that of ducks. The question arises then as to whether flamingos are ducks with long legs, or storks with webbed toes. Bird lice, Mallophaga, would appear to give the answer and one is reminded of a remark of Ernst Mayr of Harvard, when chairman at the first Symposium on Host Specificity, that 'Mallophaga are the bloodhounds that track down the phylogenies of their hosts'. *Colpocephalum* is found on storks, herons, ibises and spoonbills and is thus an indubitable parasite of Ciconiiformes. It is also parasitic on flamingos. But in addition there are on flamingos three genera *Trinoton*, *Anaticola* and *Anatoecus* which are also found on ducks and not on any other birds. The weight of the evidence would appear to be in favour of a relationship with ducks, but we must bear in mind E. Mayr's rider, that 'mallophagan bloodhounds sometimes bark up the wrong tree'. The evidence from cestodes shows no relationship whatsoever with ducks for the tapeworms of flamingos are *Amabilia lamelligera* (Owen), *Leptotaenia ischnorhyncha* (Lühe) and *Gynandrotaenia stammeri* Fuhrmann. These cestodes are monotypic and are not found in any other bird, but both *Gynandrotaenia* and *Leptotaenia* are genera of a very peculiar and clearly defined family in which the other genera are parasites of Charadriiformes, that is of waders such as plovers, oyster-catchers and avocettes. If we look to the nematodes for evidence we find that the genus *Echinura* Soloviev contains species parasitic in Anseriformes (ducks and geese), Ciconiiformes (storks), Charadriiformes (waders) and Phoenicopteriformes (flamingos) and so we are no nearer a solution to the problem.

The case of the crested seriema, *Cariama* (=*Dicholophys*) *cristata*, a running ground-bird of the pampas of Brazil, is quoted by Baer as showing both the possibilities and the limits of establishing the phylogenetic affinities of an isolated group of birds. The seriemas are the only living descendants of large carnivorous ground-birds whose remains have been found in Oligocene, Miocene and Pliocene strata in South America. *Cariama cristata* is infested with trematodes, Acanthocephali, nematodes and cestodes. The trematodes are also found in South American snake-eating eagles, while the Acanthocephali occur nowhere other than in South American fish-eating eagles. Two species of nematodes are also found in bustards in the Old World, and of the two cestodes one of them is peculiar to the seriema and the other is also a parasite in a bustard in Europe. The trematodes and Acanthocephali obviously indicate ecological specificity, for the seriemas, primarily insectivorous, feeding on large ants, are somewhat omnivorous and do include lizards and snakes in their diet. The indication of a phylogenetic relationship from the nematodes is reinforced by the knowledge that the other hosts are in the Old World, an indication which is confirmed by the cestode which is also from a bustard and from a European one. Parasitological evidence demonstrates an affinity to the bustards (Otididae), and the antiquity of these groups is attested by fossil forms from the Eocene of Europe which must be about 50 million years old and which link these two groups.

An association between a parasite and host is occasionally found which can neither be explained on ecological grounds nor on phylogenetic grounds. Such is the case of *Ophryocotyle zeylanica* Linstow, a cestode parasitic in the Ceylonese hornbill. The genus *Ophryocotyle* Friis comprises some dozen species, most of which are parasitic in Charadriiformes while two show an ecological specificity as parasites of Ciconiiformes, while the coraciiform hornbill is the exception. Since the Ceylonese hornbill is a shy bird of heavy forest, mainly frequenting tree-tops and feeding on berries and insects, it is difficult to account for its parasite *Ophryocotyle*, found elsewhere in waders or birds associated with the water. One might add that von Listow's record of this parasite is not an isolated one, for the parasite has been taken by the author on a number of occasions and a comparison with the type species *O. proteus* Friis leaves no doubt as to the identity of the genus. We know little of the intermediate hosts, there are only two known, a limpet in one species and a nereid worm in another, which does not help towards a solution, though it is possible that a full knowledge of the intermediate hosts of the other species would furnish an explanation.

BIOTIC POTENTIAL AND ENVIRONMENTAL RESISTANCE

Darwin's theory of natural selection is based on three observable facts and two deductions therefrom. The first fact is that all animals tend to multiply in geometrical progression, the second that the numbers of a given species remain more or less constant, and the deduction from this is that there is a high death-rate, which Darwin interpreted as depending on competition for survival or a struggle for existence. The third fact is that all organisms vary one from another, some variations proving advantageous in the struggle for survival and others unfavourable. The second deduction is *natural selection*, deduced from deduction 1 and fact 3. This is not the place to discuss which variations are inherited or how these variations arose, the question under consideration is the high rate of reproduction in parasites, Darwin's first fact, which one can label the biotic potential, and the factors that determine the constancy in numbers, i.e. the cause of Darwin's second fact, which one can regard as the environmental resistance.

It is commonly observed that parasites are prolific, although all parasites are not equally so, while there are many free-living forms that approach parasites in the number of offspring produced. In *Ascaris* the large round-worm of man, a statistical analysis of the concentration of eggs in faeces in relation to the number of parasites present gives the astonishing figure of 200,000 eggs per day produced by a single female worm, or 2·3 eggs per second, or in the course of its lifetime of two years a total of $1{\cdot}5 \times 10^8$ eggs. The hermaphrodite cestode *Diphyllobothrium*, by asexual reproduction in the formation of the many segments of the strobila, increases the number of its reproductive units. It may live for 15 years producing two million ova per day which is 10^{10} in its lifetime.

In many cestodes the reproductive systems are doubled in each segment thus doubling the biotic potential. That this is a mutation can be deduced from the fact that in closely related hosts there are cestodes, similar in armature of the scolex, in the arrangement of the genitalia, in the manner in which the eggs develop and so on, but differing in the duplication of the reproductive system in each segment of one of the species. Thus the anatomy of *Raillietina* of many species of pigeons and fowls is seen duplicated in *Cotugnia*. A similar duplication in *Paronia* is seen in that obtaining in *Hemiparonia*, both parasites of parrots, while the arrangement in *Diphyllobothrium*, mainly parasitic in land and marine piscivorous mammals, is duplicated in *Diplogonoporus* of whales and seals. It is quite possible that although this duplication may increase the potential it may not in fact actually increase the number of eggs produced, for the reproduction of a tapeworm has been found in some species to be directly related to the quality and quantity of the food of the host. Any significance in the duplication of the reproductive elements of one of the sexes is not obvious in *Diplop.. llus* of Charadriiformes where the male reproductive system is double and that of the female single, while in the same family, also parasitic in aquatic birds, Ardeiformes and Podicipediformes, there is separation of the sexes in *Dioecocestus*, the male worm possessing a double and the female a single reproductive system in each segment. In another and related genus, *Infula*, parasitic in Charadriiformes, which is also dioecious there is but a single reproductive system in each segment and the number of eggs produced is evidently adequate for the continuation of the species.

An additional multiplicative stage is seen in some tapeworms, more especially in Taeniidae where the larva produces a number of scolices; these can be numbered by the hundred in the case of a multiceps or coenurus, or by the million in the case of *Echinococcus*. Although *Echinococcus* is one of the smallest of tapeworms, formed of three proglottides, one immature, one mature and one gravid, yet as each new proglottis is formed the terminal gravid one is shed, and several are shed every day, each gravid proglottis containing between 400 and 800 eggs. Each egg has the potentiality of developing, in the intermediate host, into a hydatid cyst, and within each cyst endogenous budding gives rise to brood-capsules which in turn produce scolices. The appearance of the vast number of these minute scolices has earned for them the name of hydatid sand. As hydatid sand contains 4×10^5 scolices per ml and as the most common size of a hydatid contains 3 to 6 ml of hydatid sand the estimate of the potential reproduction of one *Echinococcus* is a *daily* one of $2{\cdot}5 \times 10^8$ and this excludes the possibility of exogenous budding which sometimes takes place.

Another mechanism whereby the biotic potential is increased is in polyembryony as seen in flukes or in *Gyrodactylus*. It has been estimated that the larger fluke of sheep and cattle *Fasciola hepatica* may produce in its lifetime some half a million ova. In its life-cycle a miracidium hatches from the egg and within the pulmonary chamber of *Limnaea* it can develop into a sporocyst which may divide to form two sporocysts, in each of which there develop 5 to 8 rediae. In summer each redia produces 5 to 8 daughter rediae and each of these in turn produces 15 to 20 cercariae. In winter, on the other hand, each first-generation redia produces 15 to 20 cercariae without the intervention of daughter rediae. If one assumes that egg-production by the parent fluke in the constant temperature of the liver of a homoiotherm is at the same rate in summer and winter, and that half the ova multiply, as do the summer generations, and half in the manner of the winter generations then, taking the average rediae and cercariae produced, one arrives at a potential total of $4{\cdot}2 \times 10^8$ offspring from one *Fasciola*. Those astronomical figures are not attained by all parasites. *Gyrodactylus elegans*, ectoparasite of carp and many other fishes, gives birth to one larva every four days, and lives for a fortnight or so, thus producing within its lifetime three larvae. But each larva contains three embryos one within the other, which are born at daily intervals, after which each *Gyrodactylus*, divested of its larvae, requires four days to become fully gravid with one full-term embryo containing three embryos one within the other. Accordingly, from one individual which gives birth to but three larvae in its lifetime, the cumulative effect of its method and rate of reproduction is such that, according to Bychowsky, at the end of 30 days the potential number of offspring from one individual is 2,453 individuals. The rate of reproduction varies according to the temperature and in a related form *Dactylogyrus vastator*, ectoparasitic on goldfish and other carp, the rate of development to maturity is four weeks at a temperature under 8°C, and four days where the temperature is 24°C.

The biotic potential is increased by hermaphroditism which gives an advantage in mating, for chance in any two individuals being together in one host is greater than chance when two individuals have to be of opposite sex. Chance in being able to reproduce is still greater when, as in the case of most tapeworms, the segments are protandrous, the male reproductive system developing first and then the female one, enabling fertilisation to take place between more anterior and more posterior segments of the same worm.

Where the sexes are separate there are various adaptations ensuring that male and female remain together. In the gapeworm *Syngamus* two individuals of opposite sex become permanently attached once they meet, as they do in many isopods and some copepods. A further example is that of *Diplozoon*, in which the larvae come together in pairs and unite to remain in a permanent condition of reciprocal fertilisation. In the hermaphrodite larvae of Didymozoonidea and of some parasitic Isopoda external factors determine that one sex or the other will develop, the result being that in an association of individuals one at least is always of the opposite sex to the other or others. A curious adaptation relating to the meeting of the sexes is that of some mites, seen by the way in some parasites of plants, where the smaller male matures first and seeks out the female hypopus, a pupal-like stage, and he carries her around on his back until she emerges.

These then are some of the ways in which a high biotic potential is attained, but the other side of the question relates to Darwin's second fact, the more or less constant numbers that obtain, and in particular the factors that control these, or the environmental resistance. These are the hazards with which a parasite has to contend, in which chance plays a large part although the dice are loaded in favour of the parasite. One can see, in a general way, that the greater number of hazards in a life-cycle involving several hosts is related to the biotic potential. Thus *Diphyllobothrium* with a biotic potential of 10^{10} has a free-swimming larva, the coracidium, with a limited and short life, and there are the hazards, even if it is eaten by *Cyclops*, of the *Cyclops* not being eaten by a fish or in turn the fish not being eaten by a fish-eating mammal, not to mention possible paratenic hosts. In *Parorchis acanthus* there is the chance that the larva, the miracidium, in the droppings of the sea-gull, will land on ground and not in the sea, where there is the hazard of it finding its molluscan host, the whelk, within the limited period of its short life. In the whelk the biotic potential is increased in the production by polyembryony of numerous cercariae which meet the hazards of finding and penetrating the second intermediate host, after which chance determines whether the infested second host will be eaten by the primary host. A contrasting life-cycle, one not involving any intermediate host, is that of the monogenean *Gyrodactylus elegans* which produces but three larvae in its life, but these meet fewer hazards in dissemination, passing directly from minnow to minnow or from parent stickleback to offspring.

Many parasites, once attached to their hosts or ensconced within them, have a certain security of tenure of their habitat and an adequate supply of food at their feet or around them, but having attained their goal it is not every parasite that can become established.

Further Reading

GENERAL ON PARASITISM

Baer, J.-G., *Le Parasitisme.* Masson et Cie., Paris (1946); *The Ecology of Animal Parasites.* University of Illinois Press (1952).

Caullery, M., *Parasitism and Symbiosis.* Sidgwick and Jackson, London (1952).

Dogiel, V. A., *General Parasitology.* Translation from Russian of the 3rd Edition published by Leningrad Press 1962. Oliver and Boyd, Edinburgh.

Rothschild, Miriam and Clay, Theresa, *Fleas, Flukes and Cuckoos: A Study of Bird Parasites.* Collins, London (1952).

Smyth, J. D., *Introduction to Animal Parasitology.* English Universities Press (1962).

TEXTBOOKS

Grassé, P.-P. (Editeur), *Traité de Zoologie, Tome IV*, 1er *Fascicule. Platyhelminthes etc.* Masson et Cie., Paris (1961).

Hyman, Libbie H., *The Invertebrates. Vol. II, Platyhelminthes and Rhynchocoela.* McGraw Hill Book Co, Inc., London, Toronto, New York (1951).

MONOGRAPHS OR TREATISES ON GROUPS OF PLATYHELMINTHES

Bychowsky, B. E., *Monogenetic Trematodes, their Systematics and Phylogeny.* Translated by P. C. Oustinoff. American Institute of Biological Sciences, 2000 P Street, Washington 6, D.C., U.S.A. (1957).

Dawes, Ben., *The Trematoda with special reference to British and other European forms*. Cambridge University Press, England (1946).

Fuhrmann, O., *Les Ténias des Oiseaux*. Mémoires de l'Université de Neuchâtel, Neuchâtel, Switzerland (1932).

Smyth, J. D., *The Physiology of Trematodes*. Oliver and Boyd, Edinburgh and London (1966).

Wardle, R. A. and McLeod, J. A., *The Zoology of Tapeworms*. Minnesota Press, Minneapolis, U.S.A. (1952).

BOOKS RELATING TO PARASITES OF MAN, DOMESTIC ANIMALS AND ANIMALS OF ECONOMIC IMPORTANCE

Brumpt, E., *Précis de Parasitologie*. Masson et Cie., Paris.

Cameron, T. W. M., *The Internal Parasites of Domestic Animals*. A. and C. Black Ltd., Soho Square, London.

Chandler, Asa and Clark, Read, *Introduction to Parasitology with special reference to the parasites of Man*. 10th Edition, John Wiley, New York and London.

Dogiel, V. A., Petrushevski, G. K. and Polyanski, Y. I., *Parasitology of Fishes*. Translated by Z. Kabata. Oliver and Boyd, Edinburgh and London.

Manson-Bahr, P., *Manson's Tropical Diseases*. 16th Edition. Baillière, Tindall and Cassell, London (1966).

Neveu-Lemaire, M., *Traité de Helminthologie, Médicale et Vétérinaire*. Vigot Frères, Paris.

Soulsby, E. J. L. (Editor), *Biology of Parasites, Emphasis on Veterinary Parasites*. Academic Press, New York, London.

Swellengrebel, N. H. and Sterman, M. M., *Animal Parasites in Man*. Van Nostrand, London, Toronto and New York.

REFERENCES FOR BIBLIOGRAPHY, TAXONOMY ETC.

Yamaguti, Satyu, *Systema Helminthum*. Vol. 1, *Digenetic Trematodes* (1958). Vol. 2, *Cestoda* (1959), Vol. IV, *Monogenea and Aspidocotylea* (1963). Interscience Publishers, New York, London.

SYMPOSIA AND JOURNALS

The literature on Platyhelminthes and Parasitology is very extensive and scattered through many journals among which the following are more restricted to various aspects of the subject.

Advances in Parasitology (Edited by Ben Dawes). Academic Press, London and New York.

Annales de Parasitologie humaine et comparée, Paris.

Experimental Parasitology, New York.

First Symposium on Host Specificity among Parasites of Vertebrates. Edited by the Secretary of the Symposium, Jean G. Baer, Neuchâtel, Imprimerie Paul Attinger S. A. (1957).

Helminthological Abstracts, Commonwealth Bureau, St. Albans, England.

Immunology, Blackwell Scientific Publications, Oxford, England.

Journal of Helminthology, London.

Journal of Immunology, Baltimore, U.S.A.

Journal of Parasitology, Lancaster, Pa., U.S.A.

Parasitology, Cambridge, England.

Parassitologia, Berlin.

Symposia of the British Society for Parasitology, Angela E. Taylor (Ed.), Blackwell Scientific Publications, Oxford and Edinburgh.

First Symposium (1963) *Techniques in Parasitology.*

Second Symposium (1964) *Host-Parasite Relationships in Invertebrate Hosts.*

Third Symposium (1965) *Evolution of Parasites.*

Fourth Symposium (1966) *The Pathology of Parasitic Diseases.*

Fifth Symposium (1967) *Problems of* in vitro *Culture.*

Sixth Symposium (1968) *Immunity to Parasites.*

Zentralblatt für Bakteriologie, Parasitenkunde, Infektionskrankheit und Hygiene, Jena, Stuttgart.

Zeitschrift für Parasitenkunde, Berlin.

Index

Page numbers in **bold** *denote illustrations.*